Cheap and Clean

Cheap and Clean

How Americans Think about Energy in the Age of Global Warming

Stephen Ansolabehere and David M. Konisky

The MIT Press
Cambridge, Massachusetts
London, England

MIT Press books may be purchased at special quantity discounts for business or sales promotional use. For information, please email special_sales@mitpress.mit.

This book was set in Sabon by Toppan Best-set Premedia Limited. Printed and bound in the United States of America.

Library of Congress Cataloging-in-Publication Data
Ansolabehere, Stephen.
Cheap and clean : how Americans think about energy in the age of global warming / Stephen Ansolabehere and David M. Konisky.
pages cm
Includes bibliographical references and index.
ISBN 978-0-262-02762-5 (hardcover : alk. paper) 1. Clean energy industries—United States—Public opinion. 2. Renewable energy resources—United States—Public opinion. 3. Energy policy—United States—Public opinion. 4. Global warming—United States—Public opinion. 5. Public opinion—United States. I. Title.
HD9502.5.C543U624 2014
333.790973—dc23

2013048414

10 9 8 7 6 5 4 3 2 1

To our children, Rebecca, Julia, Ben, and Will.

Contents

Acknowledgments

This project is very much a product of MIT. It emerged out of the MIT Energy Studies and the MIT Energy Initiative and fully reflects the can-do attitude of that great institution.

We are deeply grateful to Ernest Moniz and John Deutch, who provided intellectual leadership for the energy studies and initiative, and to Paul Joskow and Richard Lester, for drawing us into this important subject. All four have educated us fully about this subject. We wish to thank former MIT Presidents Susan Hockfield and Charles M. Vest and Associate Provost Philip S. Khoury for their support and leadership. Others involved in the Energy Studies and environmental research at MIT had a direct and immediate influence on this project, notably Phil Sharp, Melanie Kenderdine, Howard Herzog, Jake Jacoby, Michael Greenstone, David Reiner, Stephen Meyer, and the faculty of MIT's Department of Nuclear Engineering.

We have also received tremendous support from Harvard University and Georgetown University. We acknowledge the financial support of the Dean of the Faculty of Arts and Sciences at Harvard University, the Dean of the Graduate School of Arts and Sciences at Georgetown University, and the Georgetown Environmental Initiative. We would also like to recognize the intellectual support of the Harvard University Center on the Environment.

Any research project, such as this one, also depends on the support of a wide network of researchers and research organizations. We wish to acknowledge Knowledge Networks and YouGov, the two firms that conducted most of the surveys we designed, and the research assistance of Shyamali Choudhury, Jon Glicoes, Carly Morrison, Maxwell Palmer, and Joe Williams.

We would like to thank Clay Morgan at the MIT Press for his enthusiastic support of the book, and the skilled staff at the press for their assistance in converting the manuscript into the final product.

Finally, we wish to thank our wives, Laurie Gould and Kristen Konisky. Their continued support and perseverance makes our work possible.

1
The Energy Challenge

Energy is back. After three decades of tepid public interest and scant government attention, energy is once again squarely on the national political agenda. Washington, D.C., has seen a storm of activity on this topic in recent years, after relative silence since Jimmy Carter inhabited the White House. In 2005, George W. Bush signed a new law that provided a series of tax incentives and loan guarantees to encourage energy development of all types of sources. Two years later, President Bush signed additional legislation, requiring increased use of ethanol as a fuel additive and strengthening automobile fuel efficiency standards for the first time since 1975. Barack Obama raised these standards twice in his first term. In 2009, the House of Representatives passed the first comprehensive energy legislation since President Carter's National Energy Act, and President Obama's Stimulus Bill poured tens of billions of dollars into energy research and development. The states have taken up the cause as well. More than thirty states have passed requirements for use of renewable sources of electricity, and ten states have placed limits on carbon emissions from electricity generation.[1] At least three states have new nuclear power plants under construction with many more under active consideration, and states like Ohio, Pennsylvania, and Texas are experiencing an energy boom in natural gas. Even Hollywood has gotten into the act, with movies on such sexy topics as fracking (Matt Damon's *Promised Land*) and climate change (Al Gore's Academy Award–winning *An Inconvenient Truth*).

All of this attention is certainly deserved, because the moment is ripe for a massive change in the way we generate and use energy. The beginning of the twenty-first century marks a rare opportunity to reshape the U.S. energy future and also our future economy. The nation needs to expand its energy sector far beyond its current capacity in order to meet the needs of a growing population and economy. At the same time, we will need to replace much of our existing energy infrastructure just to keep pace. The U.S. Energy Information Administration (EIA) projects a more than 20 percent increase in electricity demand over the next twenty-five years, the retirement of 88 gigawatts of electricity generating capacity due to closures of some coal and nuclear power plants,[2] and the creation of 235 gigawatts worth of new power plants and expansions of existing power plants.[3] This new capacity is the equivalent of 120 large coal power plants,[4] 200 new nuclear power plants,[5] or as many as 400 natural gas power plants.[6] The energy industry has also witnessed just in the past decade tremendous technological innovations that will push industry away from some ways of generating and using electricity and toward others. The people of the United States must grapple with very difficult decisions about how to produce energy and where, and what benefits we as a society will gain and which costs and risks we must avoid.

The United States, however, lacks a comprehensive energy policy. We have a hodgepodge of national and state regulations developed over many years in reaction to specific crises and problems, especially the oil crisis of the 1970s. These are solutions for problems that no longer exist and that may even be counterproductive to addressing the challenges we face.[7] Paul Joskow, one of the nation's leading academic experts on the U.S. energy industry, attributes the failure of our energy policy to adapt to changing circumstances to a lack of sustained leadership in this critical area of public policy.[8] To the extent that the United States has a coherent energy policy at all owes not to laws governing the development of energy, but to laws protecting the environment. The Clean Air Act, the Clean Water Act, and other environmental laws have shaped the direction of energy use and production in the United States more than any other federal laws. These environmental acts force firms and consumers to take account of the societal harms of energy production in ways that would not otherwise be reflected by the price of electricity or the price of transportation fuel. They have made it more costly to burn coal and oil, and created at least an opportunity for less polluting forms of energy to enter the market. The powerful effect that environmental regulations have on the production and use of energy in the United States is, as we argue throughout this book, the main leverage that American society and the American government have in addressing the energy challenges we face.

The national debate over the future of energy is made more difficult still by the growing recognition that many means of energy production risk damage not only to the local environment but to the global climate. Over the past several decades, a scientific consensus has emerged that burning fossil fuels such as coal and oil on a massive scale risks increasing the Earth's average temperature by several degrees Celsius, due to the greenhouse effect produced by much higher concentrations of carbon dioxide and other greenhouse gases in the atmosphere. Economists researching this issue estimate that the costs required to reduce the carbon intensity of our economy may rival the total environmental costs associated with all other pollution regulation under the Clean Air Act.[9] The need to expand and improve the United States' energy infrastructure creates an enormous opportunity; the prospect of global warming complicates the path forward.[10]

Reconfiguring our nation's energy sector will require the engagement and guidance of the American people. Over the course of the next decades, the public will be asked repeatedly whether it will accept new power plants, electricity transmission lines, pipelines, and other infrastructure in their communities or states. We will be asked to accept, or reject, the environmental and health risks associated with new nuclear power plants, new forms of natural gas development, large scale wind "farms," and fields of solar panels. And these developments will likely have to be near population centers and in localities not accustomed to hosting energy production facilities. Local governments will need to approve locations for new energy projects. The U.S. Congress and the state legislatures will have to pass new laws that will determine what sort of energy future the nation builds—laws ranging from tax credits and loan guarantees to fuel use and efficiency requirements to funding the construction of electricity grids to environmental and safety regulations. Energy development has long been political, but the intensity of the politics of energy will only ratchet up in the coming decades.

What is America's energy future? There are no definitive answers. Technology, industry, and public policy will certainly shape the choices available. However, the horse driving the cart is surely the American people. As consumers, we act in the marketplace and demand that our energy provision have certain attributes—that it be reliable, safe, and inexpensive. As voters, we ultimately decide what the government will do. Often that choice is indirect, as when we elect our members of Congress, but at times we decide directly which laws will govern the development and use of energy. The public is one of the central forces in the broad complex system through which energy will be produced and used. Americans, as consumers and voters, demand of industry and government delivery of energy that is cheap and clean.

The American public often acts as an important constraint on what energy we use. Protests against the construction of nuclear power plants in the 1970s, for example, slowed construction at the targeted facilities and helped bring the development of that power source to a standstill. More recently, local opposition has slowed or stopped a wide variety of energy projects, from the deployment of the nuclear waste storage facility at Yucca Mountain, Nevada, to the Keystone XL Pipeline Project through the state of Nebraska to Cape Wind Energy's proposed wind farm off the coast of Massachusetts.

The American public can also point the way forward. Americans know what they want, and, when governments engage with the public, they have been able to transform the energy sector. Perhaps the most powerful example comes from the state of Texas. In the mid-1990s, eight utilities in the state of Texas decided to develop 1,000 megawatts of electricity, and they committed to do what the public wanted. To determine what the people of Texas wanted, the state Public Utility Commission sponsored a series of town meetings and phone polls, conducted over a two-year period. In the end, Texans went all in with wind power, a surprise given the state's immense reserves of oil and natural gas, and in 1999 the State Legislature adopted renewable fuels requirements that included a heavy investment in wind power.[11]

Over the past decade, Americans have increasingly become involved in making decisions about energy. Through ballot measures and legislation, the public in most states has been called on in one way or another to help shape energy development and policy. Open meetings have become a matter of course in choosing locations for energy facilities, and many utilities now give individual consumers the option to buy energy from alternative sources, such as wind and solar, albeit at a higher price.

The level of engagement of the public in the development of energy will only grow as the industry attempts to locate new facilities and governments try to set in place regulations to balance energy needs and other public interests and priorities.

Social scientists, especially political scientists and public opinion researchers, have been oddly quiet on the energy question. The redevelopment of the United States' energy sector is one of the great near-term transitions facing our country, and climate change is one of the great political and social issues facing the world. Understanding in which direction the American public wants to go is vital to understanding which energy technologies will have a ready consumer market and which technologies might face relatively less local political opposition in their deployment. Political scientists have been notably absent in the quest to understand what energy future the public wants, and why. This book seeks to break that silence.

Understanding What the Public Wants

Ten years ago, Paul Joskow and Richard Lester approached us to join a new MIT study on the future of energy. The team they were working with sought to tackle one of the most pressing questions of our generation. Given the prospect of global climate change, how must the world change the way we produce power? They had a grand vision: build three hundred new nuclear power plants in the United States by 2050. Nuclear power represents a carbon-free means of generating electricity to replace fossil fuels. Of the alternatives to coal, oil, and natural gas, it is the only one proven to offer a reliable source of power at an industrial or national scale. But, three hundred new nuclear power plants by 2050 amounts to six a year for fifty years. Given the practicalities of electricity transmission, such a plan would mean building at least one nuclear power plant near every major city in the United States, and Americans have a long history of opposing big, toxic facilities built near their communities. That struck us as a wild-eyed idea—even a fool's errand. We gladly signed on.

Joskow and Lester, however, are not wild-eyed idealists. Paul Joskow is one of the nation's most respected economists and currently president of the Sloan Foundation. His research and policy work were instrumental in the restructuring of the electricity industry in the 1980s. Richard Lester is a prominent nuclear engineer turned business professor who leads MIT's Industrial Performance Center. Spearheading the MIT Nuclear study were John Deutch, a chemist who had served in the Department of Energy under former presidents Carter and Clinton and was one of Clinton's CIA directors, and Ernest Moniz, who had served as under secretary of energy in the Clinton administration and was later appointed secretary of energy by President Obama. They are anything but naïve, ivory-tower academics. They are firmly grounded in the realities of energy and the productivity of the American economy. Like most of the scientists, engineers, and social scientists at MIT, they love the challenge of a practical problem and have a can-do attitude.

Would Americans support an aggressive expansion of nuclear power? Could a case be made, and would the American public listen? As political scientists with expertise in public opinion, we had our doubts. History was not on the MIT team's side. Ever since the partial melt down in 1979 at the Three Mile Island nuclear power plant near Harrisburg, Pennsylvania, the U.S. nuclear industry had been moribund. Although nuclear power currently provides one-fifth of U.S. electricity, the nation had not approved construction of a new nuclear plant since the 1980s. But nuclear power is an essential part of American industry and a necessary contributor to America's economic future. At the time that they approached us, in 2002, 104 nuclear power plants provided 20 percent of the nation's electricity. The expansion our colleagues envisioned would allow nuclear power to keep pace with population and economic growth in the United States and replace obsolete power plants.

Nuclear power has some important advantages. Once built, a nuclear power plant is a very reliable and efficient source of power, not subject to the intermittency problems of other "alternative energy sources," such as when the sun shines or when the wind blows. It emits almost none of the pollutants associated with coal-fired power generation, such as sulfur dioxide, nitrogen oxides, or mercury, and nuclear power plants emit almost no greenhouse gases. But nuclear power also has substantial downsides—most important, disposal of highly toxic wastes from power production over the span of centuries and prevention of the spread of nuclear fuel and technology to those who would use this technology to make weapons rather than electricity.

Joskow and Lester approached us because several advisors to the MIT nuclear study cautioned that they were in fact putting the cart before the horse. Would the American public go for such a plan? And, if not, were they wasting their breath?

Public opinion has served as a major constraint to the development of nuclear energy since the mid-1970s. Local attempts to site new nuclear plants have met with strong public opposition that has often been angry and sometimes violent. It is common for development projects to confront a reaction of not in my backyard (NIMBY), but energy projects seem particularly vulnerable given their size, visibility, and local environmental impacts. The research we performed for the MIT study "The Future of Nuclear Power" only confirmed that conclusion. A large majorityof Americans favored increasing use of natural gas, solar power, wind power, and even hydroelectric power, but not nuclear power. The nation remained split on this as a source of electricity generation, as it has been since the late 1970s.

"The Future of Nuclear Power" and our intensely rewarding interactions with engineers, scientists, and other social scientists encouraged us to take a different approach to the study of public opinion in this domain of public policy. Much of the work that came before our project focused on partisan politics and ideology, perceptions of risk, or variation in people's social and moral values.[12] Social psychologists and political scientists are instinctively driven to such explanations of people's attitudes and behaviors. What we learned from our colleagues Joskow, Lester, Deutch, Moniz, and others is that nuclear power faces some even more basic, practical liabilities. Most important, it is expensive to generate electricity using nuclear power compared with its immediate competitors, coal and natural gas. And Americans, our survey research showed, view the energy choice more through the lens of energy prices than they do through party, social values, and risk attitudes.

That insight led us to expand our survey research on public attitudes about energy beyond nuclear power, and to track public opinion about the electricity sector more broadly. How well do people understand the costs and benefits associated with different fuels? What fuels do Americans want to use and how do their preferences relate to their understanding of the attributes of those fuels? And, perhaps most important, is concern about global warming changing the way people think about energy?

These questions had, surprisingly, received little systematic study. There had been extensive survey research on nuclear power in the 1970s and 1980s, but as Three Mile Island and the antinuclear protests of the 1970s faded in people's memories, so too did interest in survey research on nuclear power and energy generally. More surprising still, while there had been considerable attention paid to the nuclear question, there had been only scatterings of surveys on solar and oil, and almost nothing was known about what people think about coal and natural gas—the two most significant fuel sources for electricity. What sort of energy do people want and why?

Americans are pragmatic. This is true of energy as it is with so many other policies. Talking with and surveying people over the past ten years, we are struck by the fact that the average person in fact has clear opinions about which energy he or she wishes to use in the future and which public policy tools are most acceptable for achieving that future. Americans want less reliance on coal and oil and expansion of solar, wind, and natural gas. There are some differences in opinions about energy between Democrats and Republicans, between rich and poor, between urban and rural denizens, but these are slight compared with the rhetoric of the day. More important, partisan and demographic differences are quite small compared to what really drives public attitudes about energy.

The real drivers of public attitudes toward energy are people's beliefs about the practical aspects of energy use. How much will reliance on a given energy source cost us, and how harmful might that energy source be to our health? Understanding of the costs and environmental harms associated with particular fuels drive public opinion about energy. And public understanding of the costs and harms and benefits associated with various energy sources also determines whether people are willing to accept regulations on industries, taxes on fuels, and other policies that shape our energy portfolio. People choose energy based on the nature of the good itself, not the associated politics. This we call the Consumer Model of energy and energy policy preferences.

The Consumer Model holds that people evaluate energy choices as goods. They value certain aspects of those goods. Those aspects, or attributes, of the goods on which people place stronger value will have greater weight in driving what the market delivers to consumers or politicians promise to voters. In the energy sector, the most important attributes are simply that energy be cheap and clean. That is, we want energy at low economic cost (low price and little inconvenience) and with little social cost (e.g., minimal health risk from pollution).

This is the central framework that has organized our exploration of public opinion about energy, and it is the framework for this book. This book is divided into three broad sections. Chapters 2 and 3 lay out the Energy Choice—what are the alternative energy sources and their economic and social costs, and what does the public prefer? Chapters 4, 5, and 6 cover the Consumer Model in greater detail. In particular, we examine how people understand the economic and social costs associated with energy and how those shape public opinion about the energy future. Chapters 7, 8, and 9 return to the Climate Challenge. Are concerns about the future climate changing how people think about energy today? And how do people's understandings of the economic and social costs of the energy they use shape their preferences about what climate policies the government pursues (if any)?

The Consumer Model is not the only framework that explains public attitudes about energy. One strain of thought is expressly political (a Political Model). Many people often sense partisan or regional divisions in energy politics. Southern states have more readily accepted nuclear power in recent years, while New England seems downright European. Those states have placed limits on CO_2 emissions, have implemented a system of cap and trade, and have renewable energy mandates. Indeed our colleagues were surprised to learn that the nuclear power surveys revealed neither significant regional differences in public preferences about what energy the nation uses nor meaningful partisan differences. Party, ideology, region, and other such factors do not carry us very far in understanding what energy the public supports and what energy the public opposes.

Another important strain of thought holds that people's preferences reflect deeper moral values and judgments (a Values Model). Discussion of climate change and the environment generally often slides quickly into moral questions. In this book, we are not concerned with what people *should* think or *should* do. We are focused on how they actually do evaluate the choices facing the nation today. Even so, the Values Model might provide an important starting point if people think of energy and environmental policy in terms of moral or ethical considerations. Such models and the roots of such models have some play in our research findings, but, again, they appear of secondary importance compared with people's concerns about the economic and social costs of energy use.

Our research fits into a wider context on survey research about the environment. An extensive research community concerned with the environment exists. In many ways, it is the yin to our yang. And we are certainly not the first researchers to examine public attitudes about energy. We have learned much from that research tradition, but our approach diverges from that taken in previous scholarship. Until the last decade or so, most public opinion research on energy was driven by specific concerns or events associated with particular types of energy. The oil shocks of the 1970s led to extensive surveying of Americans, most notably by Cambridge Energy Research Associates, about what actions they wanted to see taken in response to the "energy crisis." Similarly, the 1969 oil spill off the coast of Santa Barbara (at the time the largest spill in U.S. waters) sparked interest in Americans' attitudes about offshore drilling for oil and gas, and more generally energy extraction. And, finally, the accident ten years later at the Three Mile Island nuclear facility intensified interest among pollsters about attitudes toward nuclear power.

Scholars studying the impact of these events have advanced our understanding of how people respond to such crises, and how it shapes their overall attitudes toward the affected energy sources. Eric R. A. N. Smith's terrific book *Energy, the Environment, and Public Opinion* comprehensively examines the polling done, both nationally and among Californians, largely in response to these types of accidents, as well as other shocks to energy markets such as price spikes. Other research has considered specific energy sources from within a broader risk framework. The research of Paul Slovic and other social psychologists have advanced our understanding of why Americans have largely opposed the expansion of nuclear power for the past three decades.[13] The sense of dread and powerlessness that people associate with nuclear risks has made this way of generating electricity a hard sell for many people.

A second long-standing direction of research concerns tradeoffs between energy development and conservation or environmental protection. Questions of this type are similar in conception to those asking people whether they prioritize jobs or the environment, and essentially tap into how people trade off economic considerations for environmental ones. People's response to this question is generally a pretty good indicator of their general environmental preferences. But questions like this from Gallup, Pew, and others have been criticized for not being specific enough to inform policy choices, and for not presenting the right choices, since there need not be a trade-off at all. Setting these particular concerns aside, this question attempts to establish how people will exchange one value for another. The question seeks to do explicitly what we do implicitly: that is, assess the degree to which people are willing to trade off higher costs to achieve a cleaner energy system. Because they are often framed in terms of energy in general, however, these questions do not provide any leverage for analyzing whether and how people perceive these trade-offs for any given fuel or across fuels.

Over the past two decades, public opinion research has increasingly analyzed attitudes about energy and energy policy in the context of global warming. The extreme heat wave during the summer of 1988 brought the issue of global warming to the national public agenda, even though scientists had already been devoting significant attention to the topic for decades. Gallup administered its first poll about the "greenhouse effect" in May 1989, and there has been a steady rise in polling on the subject since. A keyword search for "global warming" or "climate change" in Roper's iPoll Databank turns up 277 questions asked during 1990s, 1,087 during the 2000s, and already 978 so far this decade.[14]

Survey research on climate change is vast. A proliferation of polls and research trying to gauge whether people care about the issue[15] has occurred, but considerably less effort has been devoted to why and what practical policies the public supports.[16] Of most relevance to our work is the degree to which people's concerns about climate change are driving their energy preferences. The work of Jon Krosnick and his colleagues provides one answer. Krosnick concludes from his surveys of the U.S. public that there is broad consensus among Americans both that global warming is occurring and that it is the result of human activity, and that Americans overwhelmingly support government actions to mitigate the problem.[17] More specifically, a survey he conducted in 2010 found that three-quarters of the public favored the government requiring businesses to reduce their greenhouse gas emissions and providing utilities tax breaks to use renewables for power generation. Yet a similar large majority also opposed specific measures such as taxes on electricity and consumption of gasoline.

Research by Anthony Leiserowitz and his colleagues contributing to the Yale Project on Climate Change Communication suggests that the American public is not quite as uniform in its beliefs about climate change. Leiserowitz has described Americans as falling into "Six Americas," ranging from individuals who are "Alarmed" about climate change to those who are "Dismissive."[18] These beliefs influence their opinions about whether action should be taken by government, businesses, and others to deal with the problem. Most Americans not surprisingly fall somewhere in between these categories, but the point is that considerable heterogeneity exists in levels of public concern and engagement with the issue, much more than what is suggested by Krosnick's work. The rich surveys regularly conducted by Leiserowitz and his colleagues also indicate significant support for the United States adopting policies that would in some fashion address the climate problem. In their September 2012 survey, for example, they find the majority of the public supporting policies ranging from funding more research into renewable fuels to regulating carbon dioxide as a pollutant to eliminating subsidies to the fossil fuel industry.

The work by Krosnick, Leiserowitz, and the many other scholars working in this area shares the same basic approach. Most of this polling asks about climate change directly, trying to pinpoint whether (and in some cases which) Americans think the problem is serious and whether it is caused by human activity, and then what, if anything, they want government and others to do about it. This is a perfectly sensible approach given the salience of the problem. But climate change opinion research does not reveal much in terms of what people think about energy itself. We learn little about which sources people want to see used, and nothing about the source of their preferences, other than the implication that such preferences are related to their concerns about global warming. This approach leaves an impression that climate change concern, more than other factors, drives people's energy preferences. But, we do not know this to be true. What is the role of perception of the price of using different energy sources? What about perceptions of other environmental harms? Are there factors as important, or even more so, than concerns about global warming? Existing studies have not and cannot answer these questions.

Answering these questions requires a different approach to measuring public opinion than has been used in the past. It requires asking new questions and figuring out ways to measure the causal effect of people's perceptions of energy costs and harms on their attitudes, not just raw correlations. We designed the MIT/Harvard Energy Surveys to gauge how people think about energy and how that in turn translates into support for and opposition to energy and climate policies.

Insights

This book is organized around four key insights.

It's a choice. What people want depends ultimately on what the alternatives are. It is less meaningful to ask people whether they support nuclear power than whether they support nuclear power more or less than competing sources of energy, such as coal and natural gas. Energy firms must make development decisions based on which of many options make economic sense. The average American thinks about energy in the same context. Yet most public opinion research has focused on specific fuels, especially nuclear power, and to a lesser extent solar and oil, isolated from comparison to alternatives. Relatively little is known about what the public thinks about coal, natural gas, and wind, and there has been almost no research that tries to ascertain what people *most* want.

Not only is energy a choice about types of fuels, it is also a choice about how society is to distribute and bear the costs of producing and using energy. Markets are very efficient at producing and delivering cheap, reliable energy. Those economic costs are central to the way we think about energy, but they must be weighed against the social costs of using energy. Burning coal and oil, for example, produces particulates and other air pollutants that cause asthma, lung disease, and other respiratory ailments. The health care costs associated with such ailments are not necessarily borne by the immediate or most intense energy users or by producers. They are borne by society broadly. Regulating or not regulating particulate emissions from coal-fired power plants, then, is a choice of how much of the social costs of energy we bear and who pays.

All energy is the same. What surprised us most in studying how people think about energy is that people think about all fuels—coal, natural gas, oil, nuclear, hydro, solar, wind—the same way. People value cheap, clean energy, and they seek energy sources to minimize both the economic and environmental costs associated with energy. We thought, given what had been written about nuclear power in the past, that environmental concerns would weigh disproportionately more when it came to nuclear power than, say, natural gas. That is not the case. As we will show in the later chapters of this book, people evaluate all fuels in terms of their environmental harms and economic costs, and they view all fuels through those same lenses. Americans want cheap and clean power, and the source is not the real issue.

Clean first. While people want energy that is both clean and cheap, most Americans want to push more strongly in the direction of cleaner energy. Those who see a fuel as very harmful are much more opposed to use of that fuel in the future than people who see a fuel as not very harmful. That difference is, in turn, much bigger than the difference in support of a fuel between those who see it as very expensive and those who see it as very cheap. In other words, the effect of environmental harm on people's attitudes is much bigger than the effect of economic costs. This reflects, we argue, the fact that most people see that electricity markets manage to keep costs relatively low and service reliable, but that private industry has been less attentive to the environmental harms. Hence, at least in the public's mind, there are likely much bigger social gains from reducing the pollution associated with energy production than from lowering economic costs.

A global disconnect. Environmental concerns weigh most heavily in people's thinking about energy, but these concerns are local not global. Global environmental concerns are quite a different story. While most people today say they are concerned about climate change, those concerns do not translate readily into attitudes about what fuels we use. Concern about global warming is at a best secondary factor in explaining attitudes about particular fuels. Most alarmingly, concern about global warming appears to be working against some of the energy sources that we would need to rely on to make near-term reductions in carbon emissions on a large scale. Specifically, concern about global warming is negatively related to support for nuclear power, even though nuclear power is the one low-carbon fuel that can operate at very large scale without any additional technological breakthroughs.

Global environmental concerns do not drive preferences about energy provision. To many political analysts in the energy field, that is a showstopper. It is not, but it does require a rethinking of how to engage the energy and climate issue in a way that the American public wants.

Think Local

Throughout our decade-long immersion in this subject, we have heard repeatedly from people that they become most concerned about energy and environment when the issues hit home, when they become localized and personalized. People find it easy to protest a nuclear power or wind development twenty miles from their homes, but there are no national protests against nuclear or wind power today. Even company executives and policy leaders tend to talk about local events and cases rather than discuss the global facts. Perhaps this is because the global issue is just too big and too far off.

Our surveys echo this sentiment. People tend to think about energy choices, energy policy, and even climate policy in terms of more immediate concerns—local environmental damage and energy prices. Smog and toxic waste have as much to do with support for limits on greenhouse gas emissions as does concern about climate change. Americans are pragmatic and think about local, immediate consequences.

For much of the past decade those on a mission to save the climate have pursued policies that target that (and only that) problem. Carbon cap and trade is designed to reduce carbon emissions by creating a market for that pollutant. It is not designed to get rid of other pollutants, such as waste water, particulate matter (PM), or mercury. Yet, people want solutions to those problems as much as (or even more than) they desire reductions in greenhouse gases in the atmosphere. However, people, as we will show, view cap and trade as narrowly about global warming.

It is time that policy makers and climate advocates listen carefully to what the public is saying. Climate-specific policies do not address the other, more pressing energy and environmental concerns that people face. The first steps on the path forward is not to devise a whole new policy regime, but to find regulations and other policy innovations that can make an immediate impact on *both* local pollutants and carbon emissions. The most popular climate policies, by far, are the ones that use existing, familiar, and proven regulatory levers to reduce greenhouse gas emissions. Such policies link global warming to other pollutants. Rather than resist that conclusion, policy advocates should take it to heart and adapt their thinking to what the public finds acceptable and what the public most wants. If policy makers and environmental advocates want to make serious inroads into the climate problem in the near term, it is time for them to listen more closely to what the public wants. Yes, most Americans express concern about global warming. But they express even deeper concerns about their own economic well-being and health.

People do not think globally and act locally (as the bumper sticker implores). Rather, people think locally, and their actions can have global repercussions. The local and immediate nature of people's understanding of energy ought to be the starting point for real change in the direction of our energy and environmental policies. That is what the public wants. For the better part of two decades, passing a comprehensive climate law has appeared a daunting task. It is daunting because the advocates are trying to change the fundamental way that people think. People do not have global perspectives and global interests at heart, especially the interests of those who will be born fifty years from now. It is far less daunting to exploit the linkages between local environmental issues and concerns and greenhouse gas emissions. We do not have to change how people think to make real and immediate progress. Rather, we have to appreciate how people think about such matters and develop new ideas for public policy that align with the costs the public will accept today and the future it wants tomorrow.

2

Energy Choices

Americans want cheap, reliable, clean energy—that seems to go without saying. But, that statement alone speaks volumes about America's energy choice. We consume electricity, transportation, and heat, not coal, nuclear, or solar power. And we want energy delivered with certain attributes—namely, energy that is inexpensive, dependable, and safe. To the typical American, even to most U.S. companies, the energy choice is not really about the fuel used. Unless they are employed or invested in a given energy sector, consumers and firms care about whether the lights and computers turn on and stay on, how much their power and fuel bills are, and whether the power they use will make them sick. In short, people value the qualities or attributes of the power they use, and not the fuel itself.

Of course, the energy industry cannot magically make electricity cheaper *and* cleaner. It faces constraints that result from the technologies available today, and many of those technologies are tied to particular fuels. Some technologies may be cleaner than others, and some technologies may be more efficient, cheaper, and more reliable. The challenge is how to balance these often competing demands. How cheap should energy be? How much pollution are we willing to tolerate? How are we to trade off the social cost of pollution for the economic value of inexpensive energy?

Society strikes a balance among these competing values in two ways—through markets and governments. For their part, energy companies operate in a competitive marketplace and try to deliver what consumers demand.[1] If consumer and firm demand is such that energy companies can make higher profits from clean energy, they will produce more clean energy. But if they can make higher profits from less clean, but perhaps more reliable, fuels, they will use those fuels. When markets cannot deliver, consumers, acting as voters, turn to government. Electricity and gasoline prices do not necessarily reflect the external costs of energy production. Markets are also not terribly sensitive to the extent to which fuels come from domestic sources or exports, whether power plants are safe, or whether ways of generating electricity or heat pollute the air and water. Markets also do not guarantee that all persons, including those who are very poor, can afford a minimum or subsistence level of heat or electricity for their homes or can afford transportation to and from work. And markets do not necessarily prevent the emergence of monopolistic energy providers. People demand legislation or regulation to compensate for failings of the market. Governments, for their part, can alter the financial incentives facing energy firms and consumers by regulating prices or levying fuel taxes, and governments can force firms to reduce pollution or require use of technologies that cause fewer externalities, even though energy prices may rise. Whatever the instruments, markets and governments are the mechanisms through which people's preferences about energy are expressed and how the public is able to balance the desire for cheaper energy against the value of a cleaner environment.

Much of this book examines how people think about the energy choice. We document through a series of surveys conducted over the past decade how people see various fuels in terms of cost, environmental harm, safety, and other attributes. We further show that people value cost and environmental harm when thinking about energy and that people tend to think about each energy source in the same way. Those perceptions, in turn, translate into preferences about what energy we use and what energy policy the United States should pursue, even as it pertains to global warming. When it comes to energy, Americans are not partisans; they are consumers.

It is instructive to begin, however, not with the choice as people see it, but with energy technology itself. The source of the trade-off between environmental harm and economic price lies with energy technology itself, not with the preferences of consumers and voters. People just try to express how they would balance two distinct attributes, given that some ways of producing electricity have more of one attribute and other ways of electricity have more of another. Ultimately, people want lower prices, a cleaner environment, and more reliable energy. Those preferences are strong. They are always present. They drive the energy sector to continually innovate. And they are ultimately the source of hope that we as a nation and a world can address the problems of global climate change embedded in our use of energy.

A Framework

Energy is supplied in our economy through complex production and distribution processes. These processes involve extraction and transport of fuels, burning or other means of using the fuels to produce electricity, the distribution of the energy to end users, and finally their use. Different fuels entail different processes, as we will discuss later in this chapter. It is natural to think about these energy delivery systems in terms of their technical features, from fuel extraction to consumer use. That is the way private and public engineering research programs are in fact organized—according to the technological aspects of fuels. Viewed from the perspective of consumers and markets, however, that is not the right framework. People don't want to think about what fuel they use because most of us put no intrinsic value on whether a specific technology or type of facility was used to generate the electricity. Rather, firms and consumers value electricity, transportation, heat, and other energy-related goods and uses, and the attributes of those goods and services. People want reliable delivery of energy, inexpensive energy, minimal harm to themselves and others, and so on. These are examples of the features of energy that people actually value and, thus, they are the attributes that will be reflected in what the market for power delivers.

Throughout this book, we examine the energy choice not so much as one fuel versus another but in terms of the *attributes* of energy provision. What are the various qualities that people value? How do they value them? And how does that valuation shape consumer preferences about fuels and voters' preferences about energy policies?

Every means of generating electricity or transportation or heat has many attributes of value to consumers. We will focus on two, which are paramount. First, there is the *price* of electricity, which is the cost to the consumer of using a given technology and fuel. Costs are determined by fuel prices, capacity constraints and storage, capital costs, and so forth.

In a competitive market, price is the marginal cost of delivering an additional amount of a given product. Energy prices are often driven by fuel prices and operating costs because many capital expenditures are sunk costs and have little effect on the ability to produce an additional amount of the good. When oil prices go up, gasoline prices follow. As natural gas prices have plummeted due to horizontal drilling and other innovations, the price of this commodity for home heating and electricity has dropped precipitously; some are even now considering the widespread use of natural gas as a fuel for transportation. Electricity prices tend to reflect the costs of the lowest priced fuel, because power plants sell to the grid, which is owned by private firms, and the grid owners will buy the lowest-priced power. That dynamic gives firms that provide power to the grid strong incentives to build power plants and stations that will have low average fuel costs, such as coal and natural gas.

Second, there are social costs—the *harms* associated with environmental pollutants emitted from the production, distribution, and use of a given fuel. Burning coal or oil releases particulates, sulfur and nitrous oxides, and other pollutants into the air. These cause smog and contribute to asthma, lung cancer, and other ailments, and air pollutants damage crops, kill forests, and contaminate water. These social costs are not reflected in the price of delivering energy to consumers, but they are borne by society through higher health care costs, damage to agriculture, and so on. These are near-term and localized costs that society pays for its current use of energy.

Such environmental harms are distinct from the potential costs and harms associated with greenhouse gases and resulting changes in the Earth's climate. These costs and risks are neither local nor near-term. They are difficult to quantify and isolate. They will hit people in developing countries harder than people in developed areas. We treat social costs as a separate attribute from local harms long into the future. And, of course, there might be other long-term risks associated with energy, such as long-term waste storage or proliferation of dangerous technologies.[2]

Many other attributes might be considered. There are *safety* concerns with every energy technology and distribution system. Catastrophic accidents may occur at mines, refineries, and power plants, and even with consumer appliances. There are problems of *reliability*. Some fuel systems work intermittently depending on conditions (such as wind, sunlight, and tides). Wind, for example, blows strongest at night in many areas and little during the day. In some areas, wind at night is so cheap that wind-generated electricity has a negative price, meaning the power company will pay you to run your air conditioner with the window open in order to get rid of the excess electricity generated. Excessive importation of fuels also causes macroeconomic and international problems. Our heavy reliance on oil imports is commonly thought to raise *security* risks, even though much of our imported oil comes from Canada, Mexico, and Venezuela.

Each of these is an important attribute of our energy production system that consumers and voters value. For many of these attributes, we would pay more today in order to achieve improvements in the near future. For simplicity, we think of energy in terms of two attributes—price and harm; that is, economic cost and social cost. Every fuel or energy technology can be characterized by these attributes. Coal, at least as we commonly use it today, provides a relatively inexpensive and reliable source of electricity, but it is quite polluting. Solar power is much more expensive than coal and not as reliable (because of electricity storage limitations), but it causes much less environmental damage and has no long-term costs in terms of global warming. One may think of technologies, then, in terms of the space of choices mapped out by the attributes. This is shown schematically in figure 2.1, which offers a rough representation of the environmental and economic costs associated with coal and solar power. Higher values along each dimension reflect *better* outcomes. So, higher values along the harm dimension mean lower environmental harms and social costs, and higher values along the price dimension mean lower prices.

The entire energy sector may be characterized in terms of a technology frontier that expresses the trade-off between harm and price using the best technologies available. This frontier expresses how much of each of the many attributes we may achieve with existing technology. There is an implicit trade-off between energy price and environmental harm (or other attributes). For any given fuel or energy delivery system, we can have a relatively low price and ignore the external or social costs of that fuel, or we can remove the pollutants during the energy production process or try to make the price reflect the social costs associated with

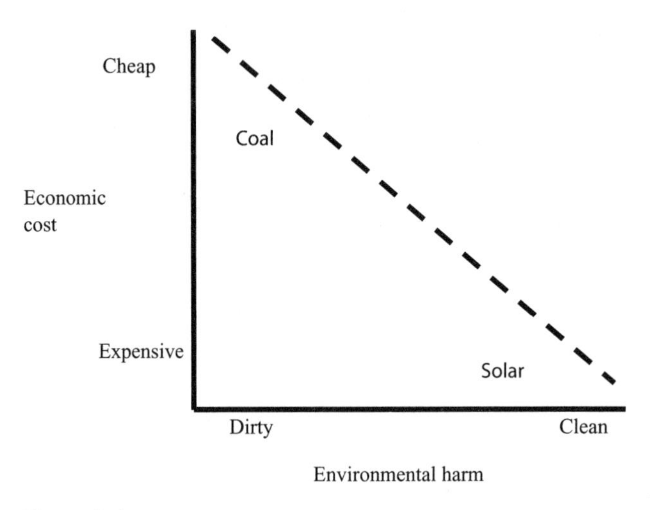

Figure 2.1
Energy attributes

its environmental and health harms. Likewise, for any mix of fuels and energy delivery systems in the energy sector, we can achieve lower prices albeit with relatively high social cost, or we can reduce the social costs of environmental harms, which will raise the price. That frontier is represented as the downward-sloping line in figure 2.1. It can represent what is possible for the entire sector, or what is possible for any given fuel.

Consider, for example, coal. Most coal-fired power plants pulverize the coal and then burn it in high-efficiency boilers. The burning splits the hydrocarbon molecules and binds those elements with oxygen, creating water (two hydrogens and an oxygen) and carbon dioxide (one carbon and two oxygens) and releasing, in the process, the energy that bound the hydrocarbon molecules together. So far, so good. This is an inexpensive and reliable way to generate electricity; it fueled the industrial revolutions in the United States and Europe. Unfortunately, the carbon escapes into the atmosphere, where it remains in higher and higher concentrations. There are also other chemical compounds and elements trapped in coal, such as mercury, sulfur, and nitrogen. Combustion releases those and creates new pollutants such as sulfur dioxide, which causes acid rain, nitrogen oxides, which contribute to smog, and fine particulates, which in high concentrations cause asthma and other lung diseases. It is possible to trap many of these pollutants and store them as waste products, but doing so adds to the cost. It is, for example, possible to reduce carbon emissions from coal with a technology called integrated gasified combined cycle coal (IGCC): first, coal is refined and turned into a gas; during that process, a significant amount of the carbon is extracted from the coal. The gas may be used to generate electricity, and the carbon can be stored deep underground in stable rock formations for thousands of years. The carbon will eventually adhere to other rock and return to a solid state. This approach removes much of the greenhouse gas associated with burning coal. It is also estimated to double the cost of coal-fired electricity. Hence, coal may be either cheap with high carbon emissions or much more expensive with lower carbon emissions.[3] IGCC makes coal more expensive (a move downward in figure 2.1) and less harmful (a move to the right). Whether such a technology is feasible economically will depend on how much more expensive and how much less harm it introduces, especially compared to alternative technologies.

The example of pulverized coal and IGCC reveals the constraints on technology in the short run and the nature of the technology frontier. Given the available technology, we face a trade-off between price and harm, between economic cost and social cost. That trade-off depends on the availability of substitutes, which is a critical difference between electricity and transportation. Many substitutes exist in the electricity market, and the United States relies on a mix of coal, natural gas, nuclear power, oil, wind, solar, and hydroelectric generators to supply its electricity. Any improvement in one of these can alter the market. Clean coal that can be produced at approximately the same price as pulverized coal today would be a game changer, as would relatively inexpensive wind and solar power. The technology frontier for electricity, then, reflects the characteristics of various fuel sources and technologies for converting those fuels into electricity. There are many alternative energy sources, and it is a fairly competitive market.

Transportation, in contrast with electricity, lacks true substitutes. Transportation is still driven almost entirely by oil. A small amount of natural gas, biomass, and electricity is used, but oil provides nearly all of our transportation fuel owing to limitations on convenience, reliability, and storage for nearly every other fuel. The technology frontier for transportation, then, is largely the technology frontier for oil. The technology frontier for electricity depends on the attributes of all of the alternative fuel systems.

There are examples throughout history of fuels being replaced with alternatives because they improved upon the attributes of price and harm. For example, wood, animal fat, and vegetable oils were staples of the heating and lighting systems of human societies until the nineteenth century. Burning wood, however, is very polluting, especially in terms of particulates, and deforestation made the fuel increasingly expensive by the beginning of the nineteenth century. With the advent of modern geology and mining, coal became very cheap and quickly replaced wood as the fuel of choice for heating and, eventually, electricity generation. Petroleum, likewise, replaced animal and vegetable fats. Wood and fat are still available technologies, but they are simply much more expensive, more polluting, and less reliable than fossil fuels. Today, we are seeing natural gas replace coal in electricity generation because horizontal drilling and fracking have increased the supply of natural gas and caused that fuel's price to fall to below that of coal. Natural gas emits many fewer pollutants regulated by the U.S. Environmental Protection Agency (EPA) and thus has smaller social costs and lower regulatory burden. Natural gas has emerged as a superior fuel to coal, and there has been a huge increase in natural gas-fired electricity generation capacity since the beginning of the 2000s.

The trade-offs embodied in the technology frontier are short-term. Technological innovations make energy production more efficient, lowering the economic cost of energy from a given fuel, lowering the social costs associated with a fuel, or both. Such innovations have the effect of pushing the technology boundary outward. For the same level of social expenditure on energy and energy-related costs, we can achieve more economic prosperity or raise our national income. Or to put it another way, for a given national income and energy budget, we can achieve the same level of productivity with less economic cost, less social cost, or both. Some technological innovations that would push outward the technology curve are fuel-specific, such as horizontal drilling and natural gas or IGCC and coal. If the price reductions offset the social costs of drilling or if the social costs of sequestering carbon offset the price of coal, then these technologies represent improvements on our existing energy technology system that would make it possible for our society to get more out of its energy sector.

In the short run, our choice is fairly static. We try to balance the costs and benefits of different fuels so as to minimize energy prices and social costs associated with energy. Over the long run, innovation pushes that frontier forward if there is sufficient demand for technologies that improve prices or human health or the global environment. Our argument is that prices and local environmental harms are the primary attributes that drive public support for using one form of energy over others, and for public policies that would make some forms of energy more widely used or less widely used. Much of the rest of the book will examine how people make that trade-off in the short term and what it means for our energy future. But, these two attributes—economic cost and social cost—clearly shape how our economy uses energy, because public demand for energy that is clean, reliable, and cheap clearly drives technological innovation in energy. That lesson is borne out in the long view of energy.

The Long View

Over the long run, demand for cheap, reliable, and clean energy has fostered innovation that has made energy provision more efficient, less costly, and less polluting. Consumers and firms value these attributes, creating incentives to innovate in order to meet that demand. That lesson emerges from three basic facts about the U.S. economy and energy use. First, since the end of World War II, the United States has steadily reduced the energy intensity of its economy. For every unit of energy used, we produce more goods, services, and national income than we did the previous year. Second, the price of energy has fallen steadily. Third, pollution has fallen, at least since 1971 when the federal government started systematic measurement. The United States has achieved these improvements, as Michael Graetz has eloquently argued, without a coherent national energy policy.[4]

Let us consider each of these facts in closer detail.

First, the American economy has become much more efficient in its use of energy. Energy consumption is commonly measured in British thermal units, or Btus. To put a Btu in context, consider cooking on a gas stove. One Btu is the amount of heat given off by the gas burner to raise the temperature of one pound of water by one degree Fahrenheit. Total U.S. energy consumption equaled approximately 98 quadrillion Btus in 2010. The U.S. economy was also valued at nearly $15 trillion. In other words, it took approximately 7,000 Btus to produce $1 worth of U.S. GDP in 2010. That, it turns out, was a tremendous improvement.

The U.S. Energy Information Administration publishes an annual inventory of energy used in the United States from the end of World War II to the present. That assessment reveals a steady and substantial improvement in the energy efficiency of the U.S economy. Immediately after World War II, the United States used over 17,000 Btus to produce $1 of GDP (in real terms). By 1970, this had fallen to less than 14,000 Btus, and in 2009 it had declined all the way to 7,000 Btus.[5] In other words, the United States uses only about 40 percent as much energy today as it did in 1950 to produce the same amount of income. This has not resulted from belt tightening or caused the collapse of the U.S. economy. On the contrary, the nation's gross domestic product has grown tremendously since 1945, and we have achieved that while using our resources more efficiently. In fact, most of that improvement in energy use has come in the relatively short time span since 1980—that is, over the decades since the last great energy crisis.

Nonetheless it is possible to be more efficient. Among the largest economies in the world, the United States is in the middle of the pack in terms of energy intensity, alongside Japan and Brazil. China currently uses approximately twice as much energy as the United States to produce a unit of GDP, but we use almost twice as much energy as the United Kingdom to produce $1 of GDP,[6] and the United States uses 15 percent more energy than all of the countries of Europe combined, which is a comparable-sized population and total GDP.

Second, the real price of electricity has fallen steadily over the past half century. Figure 2.2 shows the trend in real electricity prices paid by the average consumer from 1960 to 2010. The EIA tracks prices of electricity in various markets and calculates average, peak, and minimum prices for each year. The price that a typical consumer pays for electricity now is roughly 60 percent of what the typical consumer paid fifty years ago. At least three factors have contributed to declining prices.

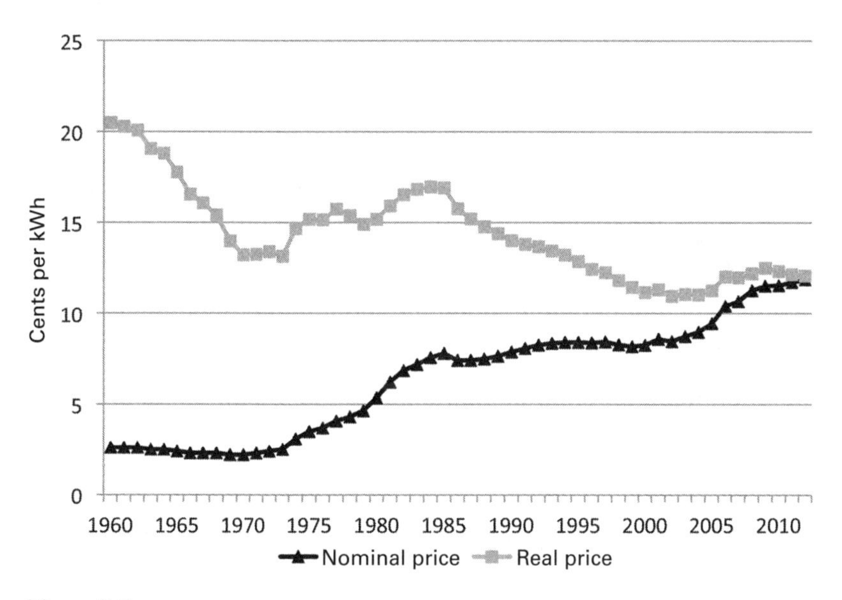

Figure 2.2
Annual average residential electricity price
Source: U.S. Energy Information Administration, *Short-Term Energy Outlook*, March 2013.

Deregulation of the energy sector produced more price competition, which has benefited consumers, as reflected in the decade-long slide in energy prices from 1985 to 1995 following the push to end public utilities' control over power generation and distribution. There have been periods of boom in power plant construction, followed by lower electricity prices, such as the one following the expansion of coal and nuclear power in the 1960s and 1970s, and the expansion of natural gas in the 1990s. And, there have been drops in fuel prices, most recently natural gas, that have lowered the marginal cost of producing electricity.

The third long-term trend is the decrease in pollution, both in absolute levels and per unit of energy used. Figure 2.3 shows trends in four key pollutants from electricity generation—particulates, sulfur oxides, nitrogen oxides, and carbon dioxide. The absolute level of pollution from the energy sector has dropped or remained constant (depending on the pollutant) since 1970, even though the total amount of energy used has increased. The amount of pollution per unit of energy or economic value has fallen dramatically since the passage of the Clean Air Act, with much

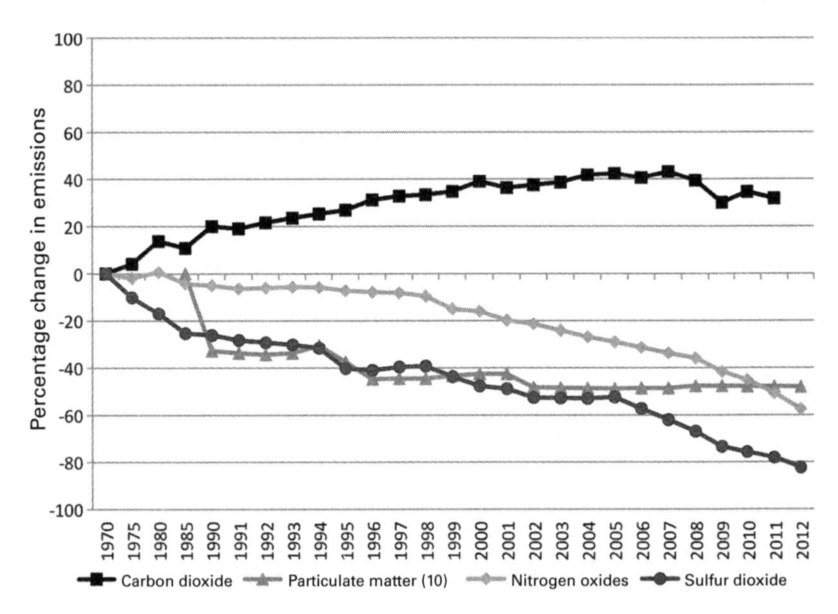

Figure 2.3
Trends in U.S. air pollutant emissions
Sources: Energy-related CO_2 emissions, U.S. Energy Information Administration, *Annual Energy Review, 2012*; SO_2, NO_x, and PM_{10} emissions from U.S. Environmental Protection Agency, *National Emissions Inventory, June 2012.*

of the decline coming after 1990, once the Clean Air Act was in full force. This is true of every pollutant except carbon, which has only recently come under the scope of the U.S. EPA.

Carbon dioxide shows a dramatically different trajectory than the other major pollutants regulated by the EPA. Total carbon dioxide emissions from the U.S. energy sector have continued to trend upward since 1970, even as other emissions have been reduced and as the energy intensity of our economy has improved. Rising emissions owe partly to the lack of regulations and partly to lack of demand. Only recently has society become aware of the potential environmental and economic consequences of carbon emissions into the atmosphere, and only since 2007 has the EPA had the legal authority to declare carbon dioxide a pollutant and to regulate it under the Clean Air Act. But, as awareness of the problems associated with carbon dioxide and other greenhouse gases has grown, it is also the case that the U.S. energy industry has reduced its carbon intensity. Since 1995 carbon dioxide emissions per unit of energy produced or per unit of GDP have fallen somewhat. There is some evidence, then, that the U.S. energy sector is already starting to move toward lower carbon intensity. The question, of course, is whether it is moving fast enough.

Technology Today

Which energy sources we use to fuel our economy matters because those fuels are tied to certain attributes. Each of the main fuels used to generate electricity has strengths and weaknesses. At one level, though, they all use the same basic technology (except for photovoltaic solar power), and how that technology operates is key to understanding the economic and social costs associated with each fuel.

Imagine a large coil of copper wire, ideal because it is highly conductive. Now imagine a large spinning magnet. Move the spinning magnet through the copper wire coil and an electric current is created, generating electricity that can be used for lighting and other work. That thought experiment, called electromagnetic induction, was the brilliant insight of British physicist Michael Faraday in the 1830s and, with it, electricity generation became commercially viable.

Nearly all power plants in the United States and the world use Faraday's method to generate electricity.[7] In a power plant, the rotation itself is done with a turbine, a cylinder with fins that are spun using steam, air, or water. Where power plants differ is in the way they spin the turbine that feeds the energy to rotate the electric generator. *Hydroelectric* power plants use water that is released through a dam on a large river to turn the turbine. This is perhaps the simplest method of power generation because it does not require more complicated energy transfers. *Fossil fuel* power plants burn the coal, oil, or natural gas to generate steam. The highly pressurized steam travels through pipes to turn the fanlike blades of a turbine. Some gas power plants operate like jet engines. The gas is burned to create heat, the heat produces pressure, and the pressure turns the turbine. *Nuclear* power, at least as it is commonly deployed in the United States, uses nuclear fission to create reactions that heat water to very high temperatures, whose steam and pressure is used to turn the turbines. *Wind* power relies on wind to move a massive propeller, which transfers the energy through a system of gears to turn a turbine. *Solar* thermal plants work by concentrating sunlight, most often with mirrors positioned in a parabolic shape, to heat a fluid such as water. The steam produced then spins a turbine in a way similar to a coal plant. Geothermal, biomass, and other power plant designs operate in analogous ways, using pressure, steam, or water to move a turbine that generates electricity. The complications in each are how do you spin a turbine, and what are the consequences of doing so?

There is one distinctive power generation method, solar power using photovoltaic (PV) cells. Photovoltaic cells consist primarily of semiconductors, such as silicon, which absorb photons (i.e., particles of light). When photons are absorbed, their energy can cause electrons in the semiconductor to be dislodged from their atoms, creating a positive charge within the cell and an electric current along the connections among the cells. This method is fairly inefficient, but it does not have the environmental problems associated with fuels and mechanisms used to turn turbines.

Almost all electricity generation, then, does the same basic thing—turn a turbine to create electromagnetic induction. These systems differ in the fuel or energy source that spins the magnet, and the fuels differ in their economic costs, their emissions of air and water pollution, their carbon emissions, their safety, and other attributes.

The advantages of fossil fuels are evident in their use. Fossil fuels are by far the most important source of power in the United States, currently accounting for about 70 percent of our electricity production, with coal being the dominant fuel. Coal-fired power accounts for about 42 percent of electricity in the United States today, natural gas contributes another 25 percent, and oil about 1 percent. Nuclear power contributes another 20 percent of the nation's electricity. Coal, natural gas, oil, and nuclear power are the old-style, dirty, and (as we will see) often disliked power sources. They account for 90 percent of our electricity. "Alternative energy," such as hydropower, wind power, and solar power, are only 10 percent of our electricity. The reason is cost.

Economic Costs

There are several ways to think about the costs of generating electricity. For most people, the price of electricity is what they see on their monthly utility bill, and they associate it with the expense of lighting, keeping their homes cool in the summer (and perhaps warm in the winter for those with electric heat), refrigerating and freezing their food, washing and drying their clothes, and powering and charging their appliances and devices. As we discussed earlier, real electricity prices for consumers have fallen steadily over the past half century. Those prices reflect the market for power produced in a complex system involving a broad range of power sources, from fossil fuels and nuclear power plants to hydroelectric dams and windmills. Today, thanks to a diverse and robust system, consumers pay roughly half as much as they did for the same amount of electricity in 1960.

Another way to think about the costs of electricity generation is in terms of the expense of generating new power from scratch—that is, constructing and bringing online a new power plant. These costs are particularly relevant for utility executives, rate regulators, investors, and political officials, since they represent the costs of adding additional capacity to the grid. The decision of how we generate power rests not with consumers but with the companies that will invest the money to build power plants.

A common way to estimate the costs of new generation from different energy sources is to calculate the "levelized" costs. Levelized costs are the present value in real dollars of the total costs of building and operating a plant over its financial life and duty cycle. Typically, these costs are presented in terms of dollars per unit of power generation, facilitating comparison across power plant types. Levelized costs include the cost of overnight capital,[8] fuel inputs, fixed and variable operations and maintenance, financing, and assumed utilization rates.[9]

As part of its 2012 energy forecast, the EIA estimated levelized cost for each of the major types of utility-scale power plants.[10] These cost assessments capture the component and total costs of bringing a power plant online in the year 2017 (the first year a plant not under construction could reasonably be brought on to the grid), taking into account differences in capacity. The EIA publishes annual national averages in levelized costs for each fuel type. It is important to mention a couple of caveats regarding these estimates. First, there can be considerable regional variation in these estimates for any given fuel due to differences in the costs of local labor, the costs and availability of fuel, and fluctuations in natural resources (i.e., wind, sunlight). Second, these estimates can change from year to year, so they reflect a snapshot at a given point in time.

These estimates are quite revealing in several regards. First, in terms of the total costs of generating power per megawatt-hour (mWh), a combined cycle natural gas plant would be the cheapest way to produce new electricity (approximately $66 per mWh), followed in order by hydro ($89 per mWh), and wind ($96 per mWh), conventional coal ($98 per mWh), and geothermal ($98 per mWh).[11] Nuclear, natural gas via a combustion turbine and biomass comprise a middle category, with costs ranging from about $111 per mWh for advanced nuclear to $128 per mWH for natural gas combustion plants. Solar-generated electricity is considerably more expensive than the other plant types, costing over $153 per mWh for solar PV and about $242 per mWh for solar thermal.

It is noteworthy that some renewable sources have levelized costs on par with some conventional energy sources. For example, onshore wind and geothermal power have roughly the same levelized costs as conventional coal and are significantly less expensive in levelized terms than combustion turbine natural gas and nuclear power. Solar PV and thermal, on the other hand, are much more expensive, costing from two to three times that of other sources.

Both consumer price and levelized cost calculations miss another sort of costs, the costs to society of generating power. Such costs might be reflected in energy prices if there are energy taxes tailored to, say, the health costs associated with smog or the environmental cleanup costs associated with oil spills. Levelized costs may also reflect such costs if environmental regulations require additional capital expenditures, say, for smokestack scrubbers, or increase operation costs. Energy taxes and regulations, though, are attempts by government to force the energy sector to reflect its social costs.

Social Costs

Social costs are hidden from consumers. They are not factored into the levelized cost of energy production; they do not appear on an electricity bill. They are, however, real. Social costs of energy production affect other parts of the economy, such as health care costs, lost workdays, and the cost of environmental cleanup. Some even argue that maintaining a military presence in the Middle East is a cost of our reliance on oil.[12] One can set aside these more remote claims, such as national defense; the environmental and health costs are bad enough.

The social costs of different energy sources vary enormously, but they primarily take the form of effects from pollution. Most of the worst pollution in terms of health effects comes from fossil fuel power plants, particularly coal, but also natural gas. Burning coal results in significant releases of sulfur oxides, nitrogen oxides, and particulates, all of which contribute to the formation of smog (ground level ozone), acid rain, and other visible forms of pollution. Breathing air in areas with high levels of smog increases the risk of cardiovascular disease, lung cancer, bronchitis, asthma, and other ailments that require medical treatment and may be fatal for some segments of the population. Fossil fuel power plants are also responsible for vast releases of toxic substances, primarily in the form of air emissions, including mercury, a neurotoxin that can cause serious health problems, but also carcinogens such as arsenic, chromium and nickel. Not only do we breathe these as airborne chemicals, but they enter the water system and we ingest them in food.[13] Because of the health and environmental impacts of these and other pollutants, coal-fired power plants have higher social costs than wind, solar, and nearly all other fuels used for electricity generation.

A 2010 report by the National Research Council estimated the social costs of U.S. energy production to be $120 billion in 2005, or about 1 percent of the nation's gross domestic product that year.[14] Most of these costs were attributable to two sectors of the economy: electricity generation and transportation. The authors considered electricity generated from all sources, but they focused their attention on monetizing the effects of air pollution mainly from coal and natural gas, since combustion of these sources to produce electricity accounted for most of the negative consequences. Almost all of these costs were in the form of increased premature mortality, elevated expected cancers, and increased incidents of asthma and other respiratory ailments attributable to air pollution resulting from combustion use of fossil fuels for transportation and power generation. The report estimated the damages from coal-generated electricity (due only to the effects of SO_2, NO_x, and PM pollution) alone to be $62 billion in 2005, more than half the social cost associated with all energy production (electricity and transportation). Natural gas accounted for another $740 million in social costs, and most of the balance was due to emissions from the transportation sector. Importantly, these damages exclude many known, but difficult to quantify, adverse health and environmental impacts from electricity production and use, including those from other pollutants, climate change, ecosystems, infrastructure, and security. For this reason, the study's authors note that their analysis underestimates the total social costs of energy.

Economists Michael Greenstone and Adam Looney provide the most comprehensive estimates to date of the total social costs of existing and emerging energy technologies.[15] The starting point for their analysis is the levelized costs of different power plants.[16] For each technology they also calculate the amounts of traditional or criteria air pollutants, such as particulates and nitrogen oxides, and the frequency and cost of various known maladies caused by those pollutants. They also calculate the amount of carbon and other greenhouse gases produced by each technology. Greenstone and Looney use the estimates from the National Research Council report for noncarbon costs and a value of $21 per ton for 2010 for carbon costs, based on the work of a U.S. government interagency working group.[17]

Not surprisingly, the inclusion of social costs produces a very different picture of the costs of generating electricity. Greenstone and Looney first estimate the total costs for existing, conventional capacity sources—coal, natural gas, and nuclear power. Coal has the highest total social costs, estimated at 8.8 cents per kWh, followed by natural gas (6.0 cents per kWh) and nuclear (2.2 cents per kWh). The components of these cost estimates are quite revealing. Nuclear power has the lowest private costs, which for existing plants means mostly variable operations and maintenance and fuel costs, and because it does not produce criteria air or CO_2 emissions, has no additional social costs (the cost of storing radioactive waste is not considered). Natural gas has the highest private costs, but only modest social costs, estimated by Greenstone and Looney to be an additional 1.2 cents per kWh. Coal presents a different story altogether. The private cost of existing coal generation is estimated to be 3.2 cents per kWh, but the impacts of both noncarbon and carbon emissions on public health and the environment add an additional 6.6 cents per kWh, making coal about one and half times more expensive than natural gas and four times more expensive than nuclear power.

Greenstone and Looney next calculate the social costs of new generation capacity, which is particularly important when thinking about the full costs of bringing new sources online to replace aging plants and to meet growing demand. The most expensive source of future electricity generation (among the set that they characterize as providing base-load) is solar PV, backed up with a natural gas combustion turbine (this is a hypothetical plant). As the least established source of large-scale generation capacity, this is not surprising, as there are many technological hurdles that remain to be overcome to make solar PV an efficient and productive way to generate electricity on a large scale. They estimate the total costs of solar PV to be 13.2 cents per kWh, all but one cent of which is attributable to the private costs of bringing a new facility online. The least expensive sources are combined cycle natural gas[18] and hydropower, followed in rough order by geothermal, nuclear, biomass, onshore wind (with a backup, natural gas combustion turbine), and coal. In general, most of the costs of each source are in the private costs, with the big exceptions being coal which has high noncarbon and carbon social costs, and some biomass, which too can have high carbon social costs.

Last, Greenstone and Looney calculate the total costs of new generation capacity from intermittent sources—onshore wind, offshore wind, solar PV, and solar thermal. These renewable technologies are estimated to have negligible noncarbon social costs and, of course, no carbon social costs. However, with the exception of the now well-established onshore wind, estimated to cost about 8 cents per kWh, the other sources are still much more expensive given their comparative novelty. They estimate the costs of offshore wind to be 19.1 cents per kWh, solar PV to be 19.5 cents per kWh, and solar thermal to be 29.7 cents per kWh. These costs will surely to come down significantly as technological innovation occurs.

Some social costs associated with fuels are hard to gauge but known to be of concern. The best example is the social cost of nuclear waste disposal. Nuclear fission produces highly radioactive spent fuel and also lower-level waste from facilities' maintenance. According to the Nuclear Energy Institute, a typical nuclear power plant generates about 20 metric tons of spent fuel per year. Adding up the waste of the 100 nuclear reactors in the United States, this comes to a total of about 2,000–2,300 metric tons of spent fuel per year.[19] Spent nuclear fuel remains highly radioactive for thousands of years, meaning that safe storage is needed to prevent radioactive releases that would present severe public health and environmental risks.

As summarized by energy experts at MIT, "the management and disposal of high-level radioactive spent fuel from the nuclear fuel cycle is one of the most intractable problems facing the nuclear power industry throughout the world."[20] Although nuclear power plants have been contributing power to the U.S. electricity grid for over forty years, the United States still does not have a long-term storage facility in place to store the spent fuel (the United States is not alone this regard). Absent a long-term storage facility, spent fuel is currently stored on-site, either in pools of water designed to keep the rods containing the spent fuel cool or dry cask storage units. While this has proven a safe short-term solution, the accident at the Fukushima Daiichi nuclear power plant in Japan raised serious questions about this strategy in the long term.

Failure to deal adequately with these unknown risks has been fatal, not to people, but to the future of the nuclear industry. Without a clear solution to the nuclear storage and safety issues, the development of nuclear power has languished since the 1980s.

Greenhouse Gases

The risks associated with global warming represent some of the most difficult social costs to calculate. They are remote, occurring many generations in the future and having more serious effects in many developing countries. There have been several important attempts to quantify such costs and risks to society, such as the *Stern Review*.[21] While the ultimate social costs remain more a matter of speculation, the contribution of energy production to the problem is not. Electricity generation from fossil fuel power plants contributes substantially to U.S. greenhouse gas emissions. Fossil fuel combustion accounts for 95 percent of all sources of CO_2 emissions, and electricity generation from coal, natural gas, and oil plants accounts for about two-fifths of this total. The balance comes from burning fossil fuels in the transportation, industrial, commercial, and residential parts of the economy.[22]

Of the 5,209 terragrams of CO_2 equivalent emitted in 2009, about 34 percent came from the combustion of coal for electricity generation, compared to 7 percent and 0.6 percent for natural gas and oil power plants, respectively. The reason for the vast differences is not simply because more coal is burned to generate electricity than any other fossil fuel. Rather, the amount of CO_2 emitted from different types of power plant varies because of differences in the carbon content of fuel sources. The combustion of natural gas produces less than half as much carbon on a per-unit-of-electricity-generated basis, as does coal. (Oil and coal are quite similar in this regard.) In addition, the specific combustion and efficiency technologies employed within a plant type can make a big difference in carbon emissions, as they do for criteria air pollutants.[23]

Increasing concentrations of CO_2, methane, chlorofluorocarbons, and other greenhouse gases has, according to most scientific research, contributed to rising atmospheric temperatures. Continuing on the same pace of emissions in absolute levels, not just per unit of energy or GDP, will raise the risk of potentially catastrophic events, such as the flooding of major cities (even entire countries such as Bangladesh), increased damage to economies through expansion of deserts and droughts, and international conflicts brought on by scarcity and the resulting migration of populations. The extent of possible climate change and the exact costs are difficult to model, because the most serious consequences are two generations or more in the future.

The Choice

Our brief review of the energy technology field allows us to reframe America's energy choice in concrete terms. The estimates offered by the National Research Council, Greenstone and Looney, and others allow us to gauge the performance of each of the fuels along the two main attributes of energy production—economic cost and social cost.[24]

Figure 2.4 plots the levelized cost of energy production for each fuel source against the social costs of each fuel. For the sake of this analysis, we combine the social costs of local environmental harms and the social costs of carbon. In these estimates, the social costs of local environmental harms far exceed the social costs of carbon. We omit oil, since it provides only 1 percent of electricity and is not viewed by technology experts

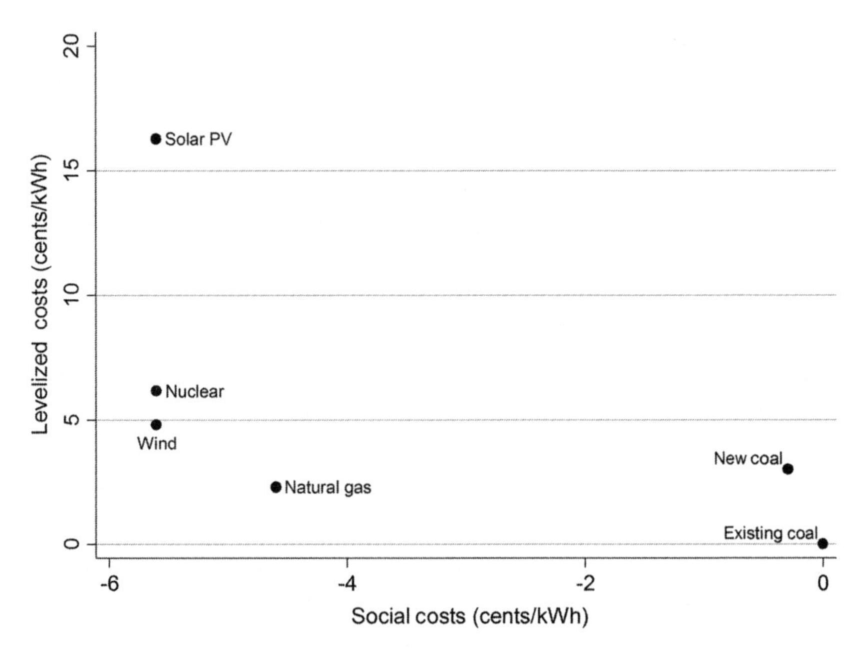

Figure 2.4
Levelized and social costs for new generation sources
Source: Data from Michael Greenstone and Adam Looney, "A Strategy for America's Energy Future: Illuminating Energy's Full Costs," The Hamilton Project, The Brookings Institution, Washington, D.C., 2011.

as a major source of electricity for the future, and we omit hydropower because its capacity is limited by the number of major rivers, and there is little room for further expansion of this form of electricity generation.

We have used as a reference point existing coal-fired power plants. These facilities account for about 40 percent of our current electricity generation. They are the least expensive (about 3 cents per kilowatt hour) but have the highest social costs from pollution and the highest social costs from carbon emissions. Along the economic dimension, the plot shows how much higher the cost of electricity production (in levelized terms) would be from building a new facility of a given type compared with the cost of electricity from existing coal plants. New natural gas facilities have a value of 2.3, which means their levelized costs are 2.3 cents per kilowatt hour higher than coal, or if we stick with existing coal rather than build new natural gas we would avoid 2.3 cents per kilowatt hour of levelized power plant cost. Of course we cannot add existing coal facilities, we can only build new ones, and new coal plants would be almost 3 cents per kilowatt hour more expensive than existing coal. The remaining three fuels are more expensive still.

Along the social cost dimension, existing coal is the worst available energy. Replacing any existing coal facility with any new form of a power plant would be a net gain in social costs. A new pulverized coal facility would improve only slightly on an existing coal facility, but a new natural gas–fired power plant would avoid roughly 5 cents per kilowatt hour in social costs compared with an existing facility. Wind, nuclear, and solar would save still more in terms of social costs.

Where does the technology frontier lie? One way to think of the technology frontier is as a budget under which consumers can choose a mix of economic costs and social costs. The sum of the social and economic costs indicates how much of our budget we have used. Existing coal is on a frontier set (by assumption) to equal 0. If there were another technology that was $5 more expensive but avoided $5 in social costs, it would be on the same frontier as existing coal. Interestingly, wind and nuclear power are very close to a net of 0. They appear to be on the same technology frontier in terms of social and economic costs as existing coal power plants.

By this account, new pulverized coal is clearly on an inferior technology curve compared with existing coal, new wind, or new nuclear power. The net value of new coal in terms of social costs and economic costs compared with existing coal is -$2.50; new coal avoids about half a cent per kilowatt hour of social costs, but is more expensive in economic terms. Interestingly, solar power appears to be on an even lower technology curve than new coal. Solar power avoids about 5 to 6 cents per kilowatt hour in social costs, but costs are more than 10 cents per kilowatt hour more expensive than existing coal per kilowatt hour of electricity.

The most intriguing fuel is natural gas. Natural gas avoids about 4.50 cents per kilowatt hour of social cost from existing coal plants, but has higher levelized costs that are about 2.25 cents per kilowatt hour more than coal (at the time of Greenstone and Looney's calculations). Hence, natural gas represents a net gain over existing coal of roughly 2.25 cents per kilowatt hour. The new technologies that have driven gas prices down are pushing the technology curve outward. Not surprisingly, there has been a rush toward natural gas in the electricity sector over the past decade. Natural gas and wind are the fastest growing segments of the power industry, and over the past ten years both have made a noticeable dent in the amount of coal used.

Today, natural gas is the clear technology winner. Expanded use of that fuel for electricity would bring substantial improvements in social costs but only modest increases in levelized costs of energy production. The net effect would be to lower the overall costs to our society of electricity use. That would bring about further improvement in GDP per unit of energy. What might change this picture? Advances in wind, solar, or nuclear technology that significantly lowered prices would bring those fuels closer to the new technology curve established by natural gas. Advances in coal technology that significantly reduce social costs without increasing prices would make coal more competitive with natural gas.

That, at least, is the view of energy today from the perspective of economic and technical experts. That view ignores the willingness of society to capture those social costs. The more pressing matter, then, is not how technologists and economists view America's energy choice, but how the people—consumers and voters—do. Do they see the social costs of coal? Do they see the economic liabilities of solar? Are they willing to set national policies to strike a balance between social and economic costs?

3

What People Want

Texas is not known for tree hugging. That made it all the more surprising when, after a multiyear planning process, eight of the state's largest utilities chose to commit to develop the next 1,000 megawatts of electricity generation using wind. More unusual still was the process by which they came to that decision.

In the summer of 1996, Central Power and Light, West Texas Utilities, Southwest Electric Power Company, El Paso Electric, Entergy Gulf States, Houston Lighting and Power, Texas Utilities, and Southwestern Public Service banded together to conduct a series of "town hall" meetings with customers. The objective of the meetings, simply enough, was to ask the people what they wanted. Over a two-year period, James Fishkin, then a professor at the University of Texas, conducted a series of meetings with relatively small groups of randomly chosen Texans. Rather than merely ask people their opinions, Fishkin brought together advocates of different energy sectors and let them make their best case to the people, and then let the people decide. Before the deliberative poll began, participants were asked their opinions about energy and their willingness to pay for various sources of electricity. Most people initially stated that they would be unwilling to pay higher electricity prices for alternative energies such as wind. By the end of the town meetings, there was a large increase in the percentage of participants who said they wanted to see investment in renewable sources as well as energy efficiency. Trailing by a large margin were fossil fuel plants. Largely on the basis of this exercise, the participating utilities expanded their investment in renewable sources of electricity generation, mostly in the form of wind power.[1] "Initially I was skeptical," said the Texas Public Utility Commission's chair Pat Wood, "but the results have been extraordinary."[2]

Today, Texas leads the nation with 10 gigawatts of wind-generated electricity, and the market for wind in Texas is booming.[3] Shell and TXU Corporation are now planning a 3-gigawatt wind farm in the Texas Panhandle. Not to be outdone, T. Boone Pickens is planning a 4-gigawatt plant in the Panhandle.[4]

The story of how Texas became a leader in wind energy is certainly unusual. State legislators, public utility commissions, and other policy makers do not normally conduct extensive polling or focus group campaigns to figure out what energy policy should be. Texas, however, is the exception that proves the general rule. Public opinion plays a central role in shaping the development of the energy sector and the rules governing the environmental consequences of energy production.

At times, the American public spurs the government to develop new energy sources and move in new directions, as was the case with wind in Texas and with the adoption of renewable energy mandates in over thirty states. At other times, the American public acts as a brake to slow or even stop development of some technologies, as was the case with nuclear power in the 1980s and 1990s. Technology, economics, finance, and many other factors certainly shape what options are feasible for the energy industry, but the development of new energy resources is also intensely political. New power plants must navigate through substantial regulatory and political hurdles related to siting, permitting, and rate setting, a process that can take up to a decade for a given plant. The public, acting as voters and consumers, often gives industry and utility commissions and state legislatures the ultimate push toward one technology over others, and the public's approval can accelerate political acceptance of power plants, pipelines, and other needed infrastructure development.

The technology options today are well-understood. But, we know very little about what Americans want when it comes to energy, and how these preferences vary across different sources. What if we conducted a national deliberative poll? What would Americans want to do? Would they favor a continued reliance on fossil fuels such as coal and natural gas, more use of nuclear power, or perhaps a shift toward renewable sources such as wind and solar power? Surprisingly, survey researchers have not conducted systematic comparisons of various energy sources. Rather, most research has focused on specific fuels, especially nuclear power; on events, such as Three Mile Island or the Exxon Valdez Oil spill; on energy providers, including companies and utilities; or on unique aspects of energy, such as opening up the Arctic National Wildlife Refuge for drilling.[5] The baseline comparisons of different fuels that Fishkin did in the Texas deliberative poll have not been measured on a national scale and in such a way that we could begin a national dialogue about where the public wants energy policy to take us.

In this chapter, we begin to paint a picture of U.S. public opinion about the energy sources used to generate electricity. Several key themes emerge. First, Americans want less reliance on traditional fossil fuels, especially coal and oil, and prefer more of what it is "new," particularly renewable sources such as wind and solar power. Second, these preferences reflect a long-standing desire among the U.S. public to shift U.S. energy consumption from dirtier to cleaner fuels, a desire that dates back many decades. In fact, to the extent to which we can track attitudes over time, we find tremendous stability in opinions about energy. Putting these pieces together, Americans express a clear dissatisfaction with the energy status quo. Americans want a different future, but they harbor some doubts about whether such a future will materialize.

Questions Left Unasked

When we joined the MIT study "The Future of Nuclear Power," our first step was to review the existing research on the subject. There was an extensive amount of research on nuclear power in the 1970s, 1980s, and early 1990s. But we were struck by the lack of systematic survey research on energy generally. Much of the work focused on specific sources of power, especially nuclear power or oil, and neglected other sources of energy.[6] In the late 1970s at the height of public discussion about energy and energy policy, a couple of polls had asked people what they most wanted the United States to use for energy. Those polls showed coal and solar power to be quite popular, especially compared to oil. As the oil crises of the 1970s receded, and the energy issue largely fell off the national radar screen, survey research on the subject dried up. Industry-sponsored polls continued to ask about particular power sources, notably nuclear power, but by the early 1990s, publicly available polling on energy had almost completely vanished. As the new century began, there was little systematic data showing what people wanted.

In a comprehensive review of the state of the research in 1980, Barbara Farhar and her colleagues at the Solar Energy Research Institute (now the National Renewable Energy Laboratory) found that a substantial proportion of the American public perceived energy to be an important national problem, with 30–40 percent consistently indicating that the country faced an energy crisis. But beyond that, Farhar and her colleagues concluded research on public attitudes about energy was a "fugitive literature."[7] At the close of a decade wracked with energy price spikes, oil embargoes, nuclear power accidents, and other sensational events and dislocations, little had been learned about what the public wanted the energy sector to be.

There is almost no survey data on public attitudes about energy before the oil crises of the 1970s. One can find miscellaneous questions asked about atomic energy and potential shortages of energy resources, but survey data of any reasonable quality and consistency is generally not available before the mid-1970s. Survey firms began to pay some attention to energy issues following the gasoline shortages and associated price spikes that resulted from the oil embargo initiated by the Organization of the Oil Exporting Countries (OPEC) in October 1973. The focus has largely been on oil supplies and exploration, strip mining for coal, oil spills, and nuclear accidents. Well into the 1990s, survey researchers remained focused on the consequences of energy extraction and transportation, such as strip mining, offshore drilling for oil and gas, and oil spills and other energy accidents.[8]

Absent from nearly all of this research by industry, academics, and public pollsters was what we view as the essential question: What energy do people want? One strain of survey research did stand out, and it is in many respects the starting point for our project. Beginning in the late 1960s, survey organizations such as Gallup and Roper asked variations on the following question: should the United States build new nuclear power plants to meet growing electricity demand? This question and close variants were posed frequently from the late 1960s through the mid-1990s, and then again more recently. The data in figure 3.1 shows responses to three sets of questions asking Americans about their support for constructing new nuclear power plants, which we compiled from a

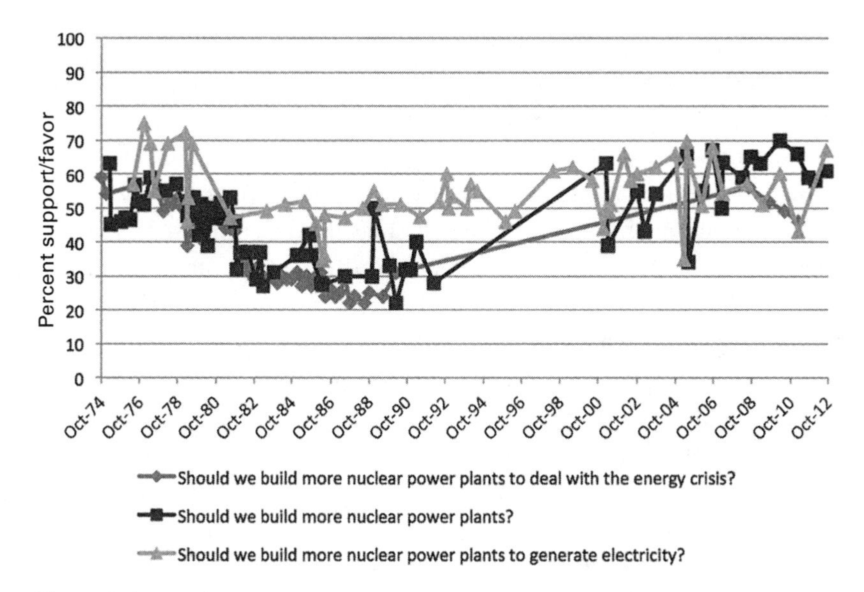

Figure 3.1
Public support for building more nuclear power plants
Source: Compiled by authors from Roper, iPOLL Database.

number of different sources. The first question, from Cambridge Research, asked respondents whether they favored the building of more nuclear power plants as an option to deal with the energy crisis. A second question asked respondents whether they favored or opposed building more nuclear power plants, without a connection to the "energy crisis." A third question, often asked in surveys conducted on behalf of the nuclear industry, asked respondents if they favored building more nuclear power plants for the purpose of generating electricity.

These data reveal that a majority of the public supported the construction of new nuclear power plants during most of the 1970s. Attitudes changed, however, after the partial core meltdown at the Three Mile Island nuclear power plant near Harrisburg, Pennsylvania, on March 28, 1979.[9] A problem with the nuclear power plant's cooling system, combined with malfunctions in the plant's warning system and operator errors, resulted in the overheating of the nuclear reactor and a partial meltdown of one of the reactor cores. While the incident resulted in only small radioactive releases, and no injuries or deaths to plant workers or the adjacent community, it dramatically changed the course of U.S. nuclear power.[10] The incident at Three Mile Island also had important consequences for public attitudes toward nuclear power. In the Cambridge Research poll immediately prior to the accident in January 1979, 50 percent of Americans favored building more power plants, while only 32 percent were opposed (the remaining 18 percent responded "don't know). Just three months later, the percentage of Americans favoring building more nuclear power plants had declined sharply to 39 percent. There was also a small, but perceptible, drop in support for nuclear power following the Chernobyl accident that occurred in the former Soviet Union in April 1986.

Public support for nuclear power has generally been on the rebound since the mid-1990s. In the most recent polls included in this time series, upward of 55–60 percent of the U.S. public has favored the construction of more nuclear power plants. Some people have posited that the increase in support for nuclear power is due to growing concerns about climate change, given that nuclear power plants do not emit greenhouse gases as part of producing electricity. However, as we discuss in some detail in chapter 5, the relationship between nuclear power and climate change attitudes is much more complicated. In fact, people that are worried about climate change tend to be *less*, not more, likely to want the United States to increase the use of nuclear power.

Beyond the nuclear question, almost no survey research exists asking about support for expansion of other fuels, such as coal or solar. A lack of consistent polling makes it difficult to say much of anything about historical attitudes toward renewable energy sources.[11] Survey researchers did infrequently ask about expansion of alternatives such as solar and wind power.[12] And on rare occasions, survey firms would ask a variant of the following question, taken from two polls conducted by Daniel Yankelovich in April 1989 and April 1991.

> In thinking about this country's future energy needs, which of these energy sources do you think we should rely on more for our future needs and which should we rely less on. . . . Nuclear power? Coal? Oil?

Both surveys showed Americans gravitated toward the alternatives, such as solar, wind, and hydropower.[13] As interesting as this question was, it did not inquire about support for each power source separately, allowing the possibility that people want to increase use of *all* energy sources. And after 1991, researchers didn't ask even this question.

MIT/Harvard Energy Surveys

To better understand how Americans think about energy, we designed a series of public opinion surveys that were administered on a regular basis from 2002 to 2013. Several of these surveys included a common battery of questions asking respondents about their attitudes about seven sources of energy used to generate electricity: coal, natural gas, nuclear, oil, hydropower, solar, and wind. These sources are used to generate about 98 percent of all electricity in the United States (the balance comes mostly from biomass and geothermal sources).[14] The questions asked respondents whether they wanted to see the use of each source increased, kept at the same level, reduced, or not used at all, as well as their perceptions of the economic costs and environmental harms associated with each source.

Among the distinguishing features of these data is that they enable us to explicitly compare attitudes across energy sources, and to do so over time. The samples taken for the MIT/Harvard Energy Surveys are repeat cross-sections of the American public, and each provides us with a snapshot of public opinion at a particular moment in time. These repeat cross-sections also enable us to detect aggregate-level changes in how the U.S. public thinks about energy, which is crucial for understanding how preferences for coal, natural gas, wind, and so on, may have changed over the last decade. Moreover, they allow us to examine how perceptions of different attributes of these energy sources, such as cost, environmental impact, safety, and security, have shifted in light of events such as the rising price of oil, large accidents in energy development (e.g., Deepwater Horizon) and power generation (e.g., Fukushima Daiichi), and growing attention to the problem of climate change.

The MIT/Harvard Energy Surveys were each administered through the Internet to different panels of adult respondents. Our choice of an Internet survey firm at the time (2002) was controversial. That part of the survey industry was new then, and was under intense criticism, largely from incumbent survey organizations that relied on random digit dialing phone polls. But the beginning of the 2000s saw a rapid acceptance of Internet surveys as a valid and cost-effective alternative to more traditional modes such as phone, mail, and in-person (face-to-face) surveys. Subsequent studies of these different approaches have not found meaningful differences among surveys conducted in these different modes.[15] The growth in the accessibility of the Internet makes web-based polling a good alternative.

The core set of surveys we analyze in this book were conducted in 2002, 2003, 2007, 2008, and 2011. The 2002, 2003, 2007, and 2008 surveys were sponsored by MIT and administered by Knowledge Networks (KN).[16] KN fields Internet surveys to its panel of potential respondents, which are recruited through random digit dialing and then contacted to participate in surveys through email. Internet access is provided for recruited panel members without it for free in exchange for completing surveys to improve the representativeness of the panel. Each of the surveys KN conducts comes from a random sample of its panel.

The 2011 survey was sponsored by Harvard and carried out as part of the 2011 Cooperative Congressional Election Study (CCES), administered by YouGov/Polimetrix (YG). The CCES survey is a collaborative project that combines the resources and expertise of scholars from universities around the country to create a large, nationally representative survey. YG uses a matched random sample methodology to generate its survey samples; it develops a target population from general population studies and then draws a random set of respondents from this target population to create a "target sample." Then, using a matching algorithm, the firm selects potential respondents from its pool of opt-in participants who match the target sample.[17]

Throughout the book, we will also report results from other original surveys that we designed in interim years, which focused on related topics. For example, surveys fielded in 2003 and 2006 focused on climate change and carbon capture and storage, a 2009 survey concentrated on nuclear power, a 2010 survey on climate change policy options, and a 2013 survey on energy costs and people's willingness to pay to address environmental harms. Table A3.1 in the appendix presents a complete list of the original surveys examined in the book.

Measuring What People Want

When we launched the MIT/Harvard Energy Surveys, only two studies truly fit what we sought to measure and broke the mold of prior research.

Jon Krosnick of Stanford University had conducted a pilot study for Resources for the Future (RFF) that asked people's attitudes about each energy source separately. That survey echoed what Fishkin had done on a smaller scale in Texas. The insight we gained from the Fishkin and Krosnick studies was that meaningful measures of public attitudes about energy use only emerge upon comparing all the feasible sources of generating power and upon digging more deeply into why people make the choices they do.

With the Fishkin and Krosnick studies as a foundation, the MIT/Harvard Energy Surveys sought to establish a national measure of what energy people want to use and to create a baseline of comparison for future survey researchers. In designing the surveys, we also sought to fix what we saw as a major weakness of most prior survey research—the lack of head-to-head comparisons of energy sources. Much survey research tries to grab a headline rather than answer a question. Media polls often ask about events, such as oil price spikes, nuclear accidents, or oil spills, and the poll numbers are used to add weight to stories about the crisis following those events. Making public policy, however, requires assessing the alternatives. Over the next two decades, over half the nuclear fleet in the United States and 20 percent of the coal power plants will come to the end of their planned lifespan. What energy sources should be brought online to replace that capacity? Should we build an equivalent amount of nuclear and coal power? And over the next twenty-five years, the United States will have to build 235 gigawatts of new electricity-generating capacity to meet growing demand. That is the equivalent of 200 nuclear power plants and nearly twice as many conventional natural gas power plants. Those new power plants will have to be licensed, and suitable locations (not far from major metropolitan areas) must be found. That will require the assent of elected officials and sometimes the public itself. What should be done to meet this growing demand?

The MIT/Harvard Energy Surveys took these questions to the public directly. To gauge what energy sources Americans want to use in the future, we included the following question on the 2002, 2007, 2008, and 2011 surveys:

Consumers, such as you, have more and more say in how electricity is produced in the United States.

To make electricity to meet the country's needs over the next 25 years, new power plants will have to be built. Companies and government agencies need to start planning today. How should we meet this demand? For each power source indicate whether you feel the U.S. should increase or reduce its use, or not use it at all.

The question then addressed seven different power sources: coal, natural gas, oil, nuclear power, hydropower (i.e., dams), wind, and solar. For each power source, respondents could choose "Increase a lot," "Increase somewhat," "Keep the same," "Reduce somewhat," "Reduce a lot," and "Not use at all."[18]

Figure 3.2 displays the responses to this question for each of the seven energy sources in each of the four surveys. For clarity of presentation, we collapse the "Increase a lot" and "Increase somewhat" categories and the "Reduce somewhat" and "Reduce a lot" categories. The fuel sources divide neatly into three categories, according to the distribution or division of public preferences. Figure 3.2 shows quite clearly where the American public wants to go.

The results are stark and consistent with Fishkin's findings in Texas, even without engaging in the deliberative process. Americans want to move away from fossil fuels, and they readily embrace solar and wind. Later, in chapter 6, we examine what deliberation about the prices and environmental consequences of these fuels might lead to, but before we get there, it is important to establish a baseline. What do people want the energy future to be like?

First, Americans want less reliance on coal and oil. Coal and oil fueled the Industrial Revolution in the United States, and they remain the workhorses of American energy production and the U.S. economy. Coal provides more than 40 percent of U.S. electricity. Oil accounts for almost all transportation fuel (though electric cars could change that), but just about 1 percent of electricity. But Americans want to get off these fuels. The first panel of figure 3.2 shows the distribution of public attitudes about coal and oil. A majority of Americans want to reduce future use of both of these fuels. The number of Americans who want to reduce the use of coal and oil outnumber those who want to increase use of these fuels by a margin of more than 2 to 1.

Second, Americans want to keep natural gas and nuclear power as they are, or increase them somewhat. Natural gas and nuclear power

a

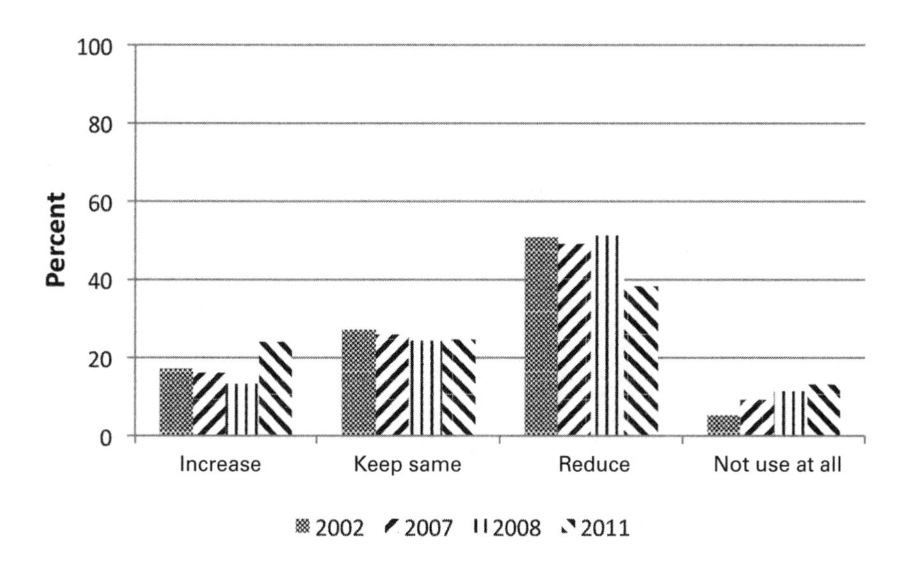

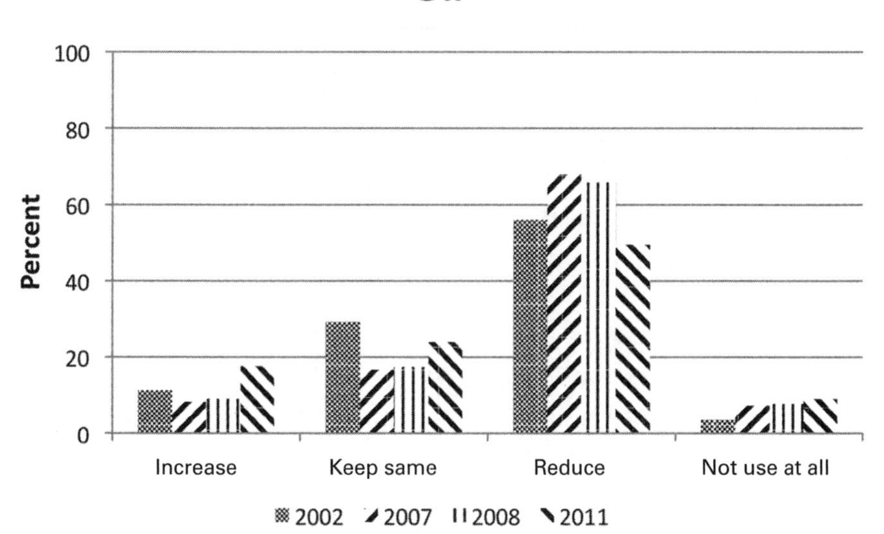

Figure 3.2a
Preferences for sources of future energy: coal and oil

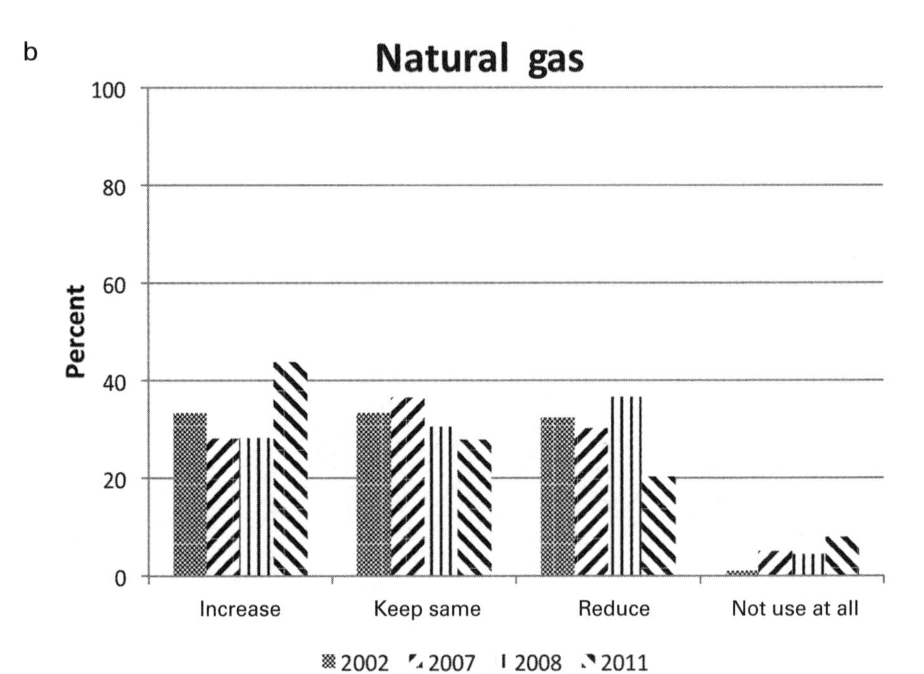

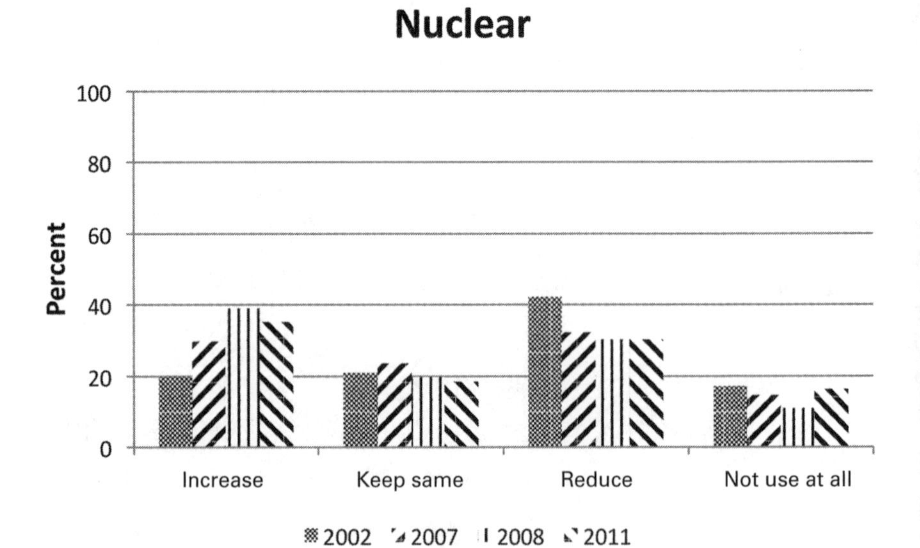

Figure 3.2b
Preferences for sources of future energy (cont.): natural gas and nuclear

c

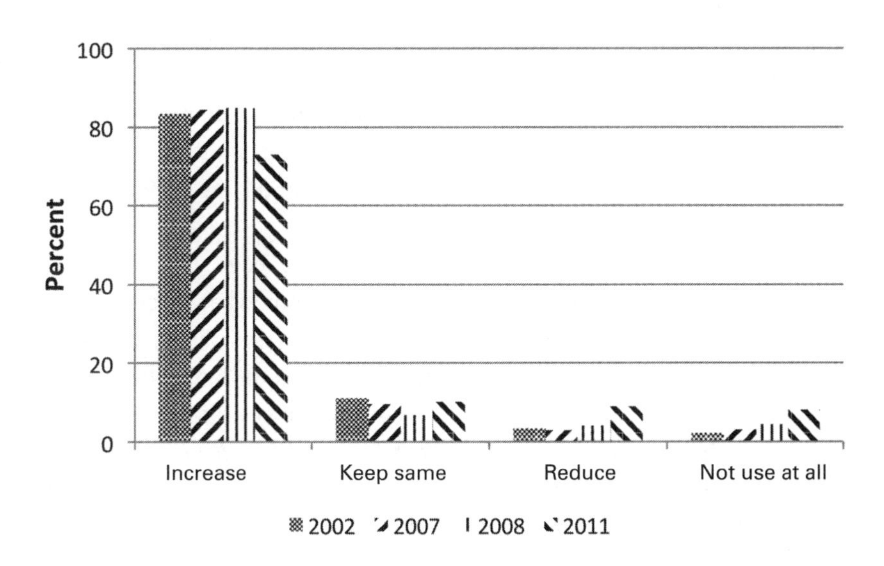

Wind

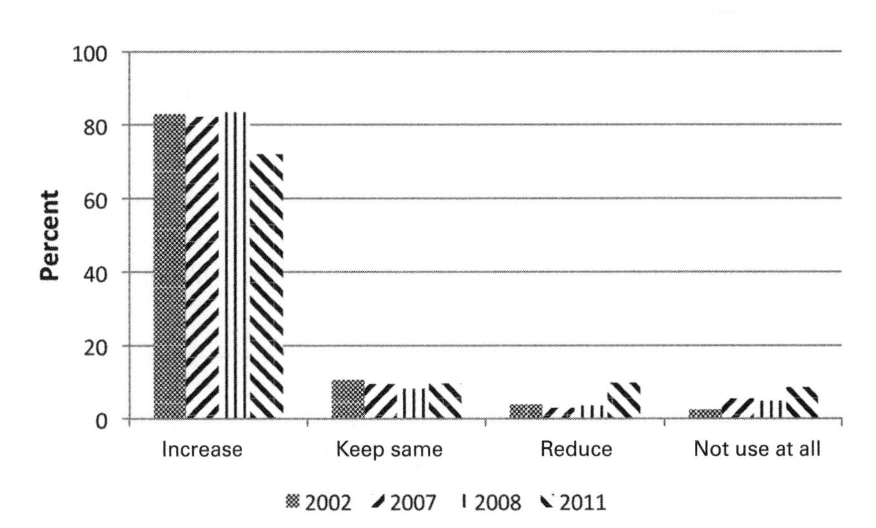

Figure 3.2c
Preferences for sources of future energy (cont.): solar and wind

have served significant supporting roles in energy production. Each accounts for about 20 percent of electricity generation, and combined they rival coal in the electricity sector. Natural gas is experiencing a revival owing to technological innovations and falling prices, and industry has shown renewed interest in nuclear power with the licensing of several new nuclear power plants. Although public support for expansion of these fuels is somewhat divided, natural gas and nuclear power today are more popular than either coal or oil, and support for them has generally trended up over the past decade.

Between the two fuels, natural gas has the edge. The number of people who want to increase natural gas is somewhat larger than the number who wants to decrease it. In 2011, there was a sharp uptick in support for expansion of natural gas. As we see in later chapters, this reflects falling prices, an important lesson for any sector of the energy industry. Nuclear power is somewhat less popular. Almost 40 percent want to expand this power source, a third want to reduce it, and another 15–20 percent do not want to use it at all. Of all the energy sources, nuclear power shows the least consensus. It has a substantial number of supporters, a somewhat larger number of detractors, and the most intense opposition.

The pattern of responses suggests an increase in support for both natural gas and nuclear power over the last decade; comparing the 2002 and 2011 responses, about 15 percent more of the U.S. public indicated it would like to see the United States use more nuclear energy to generate electricity, while about 10 percent fewer Americans wanted to see its use reduced somewhat or a lot. This is notable given that the 2011 survey was administered six months after the accident at the Fukushima Daiichi power plant.

Third, Americans want to substantially expand the use of alternative fuels, especially solar and wind. As shown in the third panel, a vast majority of Americans want to increase the use of these fuels, and only a small number want to reduce them. The results for solar and wind are especially striking. By a margin of 6 to 1, supporters outnumber detractors.

One might read these results from the perspective of firms or from the perspective of policy makers, such as public utility commissions or state legislatures. From the perspective of firms, it is clear that there is enormous unmet demand for solar and wind power. The challenge for industry is how to capture that unmet demand. From the perspective of policy makers, the question is how to push in the direction of these fuel sources, and how to reduce traditional fossil fuels, oil and coal. It helps that natural gas has begun to replace coal, as natural gas prices have fallen steeply in recent years and federal air pollution regulations have increased the cost of operating coal-fired power plants. Many states have responded to public demand for solar and wind by adopting Renewable Portfolio Standards, which require that covered utilities provide a minimum amount of a state's electricity (typically 10, 15 or 20 percent) from renewable sources.

Another way to view these data is in terms of energy use overall. Some people might want to reduce all fuels, so as to reduce energy consumption. Others may want to expand all energy sources. President Obama's energy policy has been described as "all of the above"—aggressive expansion of oil and natural gas production in the United States combined with heavy investment in solar, wind, and other alternative fuels and even expansion of nuclear power. How many want an "all of the above" approach?

Table 3.1 presents aggregate opinions about different combinations of energy sources in each of the surveys. Specifically, for each respondent we counted how many energy sources they wanted to increase and how many they wanted to decrease. We further divided the fuels into fossil fuels and alternative energies.

Very few Americans want an "all of the above" approach. Across the four surveys, the highest percentage of people who wanted to increase use of all fuels topped out at 4 percent (in 2011). The picture is not much different when looking at the proportion of the public that wants to increase the use of all fossil fuels or fossil fuels plus nuclear power, although there was some movement in this direction in the 2011 sample. Still, only about 10 percent of the population expressed this preference. There is much more consistency in attitudes across the renewable technologies, with almost 40 percent or more of the respondents in each survey indicating that they would like to see the United States rely more on all renewable sources in the coming years.

The bottom portion of table 3.1 shows similar numbers, but this time in the case of people wanting to decrease different fuel combinations. In

Table 3.1
Energy use preferences across sources

	"All of the above"			
	2002	2007	2008	2011
All fuels	3%	1%	2%	4%
Fossil fuels (coal, natural gas, and oil)	5	3	3	12
Fossil fuels and nuclear	3	2	2	10
Renewable fuels (hydro, solar, and wind)	56	41	48	39
	"Conservation"			
	2002	2007	2008	2011
All fuels	8%	7%	9%	10%
Fossil fuels (coal, natural gas, and oil)	56	61	64	46
Fossil fuels and nuclear	47	43	41	36
Renewable fuels (hydro, solar, and wind)	10	9	10	17
Sample size	452	615	623	1022

Source: MIT/Harvard Energy Surveys.
Note: Cells are number indicating preference for increased or decreased use of sources.

terms of all sources, 7 to 10 percent of the public wants to see the United States use less of all seven of the sources asked about in our surveys. That is, at most, about a tenth of the U.S. population wants to see the country use less of everything, reflecting a set of opinions consistent with wanting more conservation rather than energy expansion. The numbers grow considerably when considering fossil fuels and fossil fuels plus nuclear power. In all but the 2011 sample, a clear majority wanted to see less use of fossil fuels, and about two-fifths of the population wants to see less use of fossil fuels and nuclear power. A much smaller percentage of the population expresses a desire to see a decline in the future use of all renewable sources; about 10 percent of the population in the 2002, 2007, and 2008 surveys, and 17 percent in the 2011 survey.

Americans, then, are not particularly conservation-minded, nor have they rallied around the "all of the above" approach. Rather, a majority of the public supports reducing all fossil fuels and increasing all alternative energy. The public expresses the strongest opposition to coal and oil.

While there has been some movement toward increasing support for natural gas and nuclear power, the favorability levels of these fuels pale in comparison to solar and wind power.

Not in My Backyard Attitudes

Another way to gauge public attitudes toward energy is to measure what people do not want. Some of the great conflicts over energy in the United States have arisen not over national policy but over locating a power plant or other facility in a community or state. Communities fought hard to stop Seabrook, Pilgrim, Diablo Canyon, and other nuclear power plants in the 1970s and 1980s. Those wounds are still fresh, on both sides. The Yucca Mountain Nuclear Waste Repository was put on ice because of opposition from Nevada's Republican governor and Democratic senators. And David Heineman, Republican governor of Nebraska, opposed the Keystone XL Pipeline until an alternate route could be found.[19]

Concerns about various types of land uses, ranging from prisons and airports to landfills and hazardous waste facilities, have led to tumultuous local politics, often energizing individuals and communities into political action. Not in my backyard or NIMBY attitudes are often most intense when it comes to large industrial sites. There are many reasons why people may not want to live near large industrial sites, be they power plants, oil refineries, or factories. Large facilities such as these are often seen as eyesores, reducing the aesthetic quality of the landscape, and often resulting in nuisances such as noise and odor pollution. Further, recent work from Lucas Davis found lower housing values and rents in neighborhoods located near power plants, indicating that these disamenities are reflected in local property values.[20]

NIMBY attitudes also reflect fears about possible environmental and health impacts. Power plants are large complexes, often with massive smokestacks or cooling towers, and people living near them often worry about breathing the local air and consuming local drinking water. Whether these fears are justified is almost beside the point; the perception of risks in this context is as important as the actual risks.[21]

Opposition to local development of any form is often intense and so common that it seems to be a virtual constant of modern political life.

Any development action will be met with an equal and opposite anti-development reaction. However, not all NIMBYism is the same. Some facilities seem to evoke less strong reactions than do other facilities, and not all people react the same way.[22] From our perspective, NIMBY reactions provide another, quite concrete way to gauge public attitudes toward different sources of energy.

The questions on NIMBY we included in our surveys adopted a similar comparative approach we used for measuring perceptions of costs and harms. This allows us to investigate whether people are more or less concerned about living near particular kinds of facilities. Specifically, we asked respondents to answer the following question:

> To meet new electricity demand, utilities will have to build additional power plants. How would you feel if a new natural gas fired power plant were built within 25 miles of your home?

The response options were "Strongly oppose," Somewhat oppose," "Somewhat support," and "Strongly support." The question was then repeated for a coal-fired power plant, nuclear power plants, and a large wind power facility (one hundred 250-foot towers) (the survey did not ask about wind in 2002).

Across multiple years of our surveys, we find strong evidence that (1) NIMBY attitudes exist toward power plants; (2) these attitudes vary in their intensity by power plant type; and (3) opposition to all types of plant waned some over the course of the last decade. A summary of the responses are presented in table 3.2. As with general attitudes about various energy sources, a majority of people are strongly opposed to coal and nuclear power plants, but willing to accept a wind farm nearby. The public is divided about locating a natural gas power plant nearby.

An overwhelming majority of the public opposes a new nuclear or coal-fired power plant being built within twenty-five miles of their home. In the case of nuclear power, a majority of the public "Strongly opposed" it in each survey, while another 20 percent or so "Somewhat opposed" it. This type of opposition to living near a nuclear power plant is nothing new. Surveys from Gallup, Cambridge, and other survey organizations since the mid-1970s have found similar opposition,[23] and scholars have often credited public opposition and protests as being a large factor in the slowing down of the nuclear sector.[24] Opposition to a coal plant was

Table 3.2
Not-in-my-backyard attitudes toward different types of power plants

	2002	2007	2008	2011
Natural gas				
Strongly oppose	23.6%	21.2%	25.7%	13.0%
Somewhat oppose	30.7	32.8	32.3	26.6
Somewhat support	36.7	41.6	38.3	42.2
Strongly support	9.02	4.38	3.70	18.2
Coal				
Strongly oppose	43.7	42.6	45.0	34.0
Somewhat oppose	35.6	34.8	32.1	28.6
Somewhat support	17.3	19.9	20.2	26.2
Strongly support	3.47	2.72	2.65	11.1
Nuclear				
Strongly oppose	65.8	57.4	55.3	49.0
Somewhat oppose	18.7	21.1	21.2	20.9
Somewhat support	11.4	16.8	18.5	19.4
Strongly support	4.17	4.74	4.99	10.7
Wind				
Strongly oppose	–	7.23	11.2	12.4
Somewhat oppose	–	17.8	14.3	13.6
Somewhat support	–	47.9	48.8	38.6
Strongly support	–	27.1	25.7	35.4

Source: MIT/Harvard Energy Surveys.

almost as intense. Across the four surveys, opposition ranged from 63 percent to 79 percent.

It is important to emphasize the degree to which people are opposed to living in close proximity to a new nuclear or coal-fired power plant. As discussed earlier in the chapter, about 20 percent of the population (plus or minus about 5 percent depending on the survey year) say they want the United States to use more coal for electricity production, and about 30 percent (plus or minus about 10 percent depending on the survey year) say the same about nuclear power. Yet, most of these respondents indicate that they are themselves opposed to living within twenty-five miles of a new power plant using these fuels.[25] In other words, even among people that are conceptually in favor of using more coal and nuclear power, there is a strong undercurrent of NIMBYism.

At the other extreme, the public expressed strong support for living near a wind farm. Wind was asked about in the 2007, 2008, and 2011 surveys, and in each, three-quarters of Americans indicated support for living near such a facility, with about 25–30 percent expressing strong support. This tremendous acceptance of wind is consistent with the huge majority who believes that wind power does not generate significant environmental impacts. It is an open question whether most Americans have seen or can visualize a wind farm of the scale asked about in the question. Wind farms tend to be located in sparsely populated areas such as the plains of the Midwest and West Texas. Surveys done in some European countries have found that people living close to wind turbines are among the most supportive,[26] but we cannot say whether this dynamic exists among Americans.

Natural gas once again sits in the middle. In the first three surveys, a solid majority of the U.S. public said that they would be opposed to a new natural gas–fired power plant being sited within twenty-five miles of their home. The level of resistance, however, was considerably less intense than the case for a nuclear or coal plant; less than 25 percent of Americans said that they would oppose it strongly. Moreover, the 2011 survey suggests a reversal of attitudes. In this survey, a strong majority (about 60 percent) indicated strong support toward the local siting of a natural gas power plant.

The softening of opposition to a natural gas power plant seems to reflect a more general pattern. Across each of the conventional sources asked about—nuclear, coal, and natural gas—NIMBY concerns receded. A comparison of responses from 2002 to 2011 revealed that opposition to living near one of these facilities dropped by about 15 percent for each source. This change is consistent with a growing prioritization of energy development and a simultaneous decline in perceptions of the environmental harms associated with energy use, which we discuss in greater detail in chapter 4. In other published work, we have examined the factors driving energy NIMBYism, finding that perceptions of a fuel's environmental harms and costs are the important factors, as well as more general risk perception (risk-taking v. risk-averse), and some other demographics.[27]

A final point regarding NIMBY attitudes is worth noting. There is a sizeable proportion of the U.S. public that is opposed to everything. In the 2002, 2007, 2008 surveys, about 47 percent of the public did want a new conventional power source built in their communities, with about a third of these individuals being strongly opposed to them. Even in the 2007 and 2008 surveys with wind included, almost 20 percent of the public did want any of the new power plants located in their areas. Consistent with what we noted earlier, however, the degree of this opposition softened in the 2011 survey. Just 10 percent of the public opposed everything (including a wind power facility), although still a third opposed all conventional sources.

Who?

What people want in the energy future also involves who they want and trust to lead that change. During the 1980s and 1990s the energy sector in the United States underwent a massive change, as many local public utilities divested themselves from ownership and development of power plants, and private firms took over that responsibility. Energy markets deregulated in order to gain greater economic efficiencies and lower prices. That transformation made great sense economically, but it has a substantial political downside. Americans distrust energy companies and they blame them for many of the failings in the energy sector.

Consider for a moment the case of oil price shocks. A recent a review of energy surveys compiled by Toby Bolsen and Fay Lomax Cook showed that people tend to hold energy companies culpable for price spikes. Regardless of the specifics of the situation, Americans tend to attribute the greatest amount of blame to oil companies, more so than to oil exporting nations, the sitting presidential administration, Congress, environmental regulations, and consumers.[28] MIT historian Meg Jacobs argues that this was even true during the 1973 oil embargo, when OPEC (and Egypt and Syria) stopped supplying petroleum to countries that sided with Israel in the Yom Kippur War, and colluded to effectuate a fourfold increase in world oil prices. Americans did not blame

OPEC for the resulting crisis. Instead, Jacobs says, "They believed it was a conspiracy perpetrated by big oil to reap high profits, and they also blamed government."[29]

This type of dynamic recently repeated itself in the spring of 2011, when gasoline prices rose to over $4 a gallon in many parts of the United States. Americans again believed that oil companies were most at fault. According to a May 2011 survey conducted for CNN by Opinion Research Corporation, 61 percent of Americans thought that oil companies deserved "a great deal of blame," 27 percent some blame, and only 12 percent "not much blame" or "no blame at all." Perhaps tapping into persistent distrust of the financial industry, 59 percent attributed a great deal of blame to speculators and other oil traders. By comparison, 40 percent blamed the political and civil unrest in the Middle East, 25 percent the policies of the Obama administration, and 25 percent the driving habits of Americans. In the same CNN poll, 77 percent of Americans agreed with the statement that "oil companies as a whole are making too much profit." These recent events reveal that Americans are quick to blame energy companies for higher prices at the pump.

A series of surveys conducted by Cambridge Reports from 1979 to 1995 and Gallup from 2001 to 2012 enable us to consider opinions about energy companies and electric utilities over time. Here are the two questions:

Cambridge Reports: Now, I'm going to read you a list of various institutions or types of industries. After each one, I'd like you to tell me whether you have a very favorable, somewhat favorable, somewhat unfavorable, or very unfavorable opinion of each.) . . . [industry name]?

Gallup: (For each of the following business sectors in the United States, please say whether your overall view of it is very positive, somewhat positive, somewhat negative or very negative.) How about . . . [industry name]?

Cambridge Reports included oil, nuclear power, and electric utilities among the set of industries in their surveys, while Gallup has included oil and gas and electric and gas utilities in their surveys.

Figure 3.3 shows the net favorable rating for each of the energy sectors, compared to the average for the all of the industries asked about in that year. The rating is computed by summing the favorable (positive) values and the unfavorable (negative) ratings, and then subtracting the latter from the former. Thus, positive values indicate a net positive

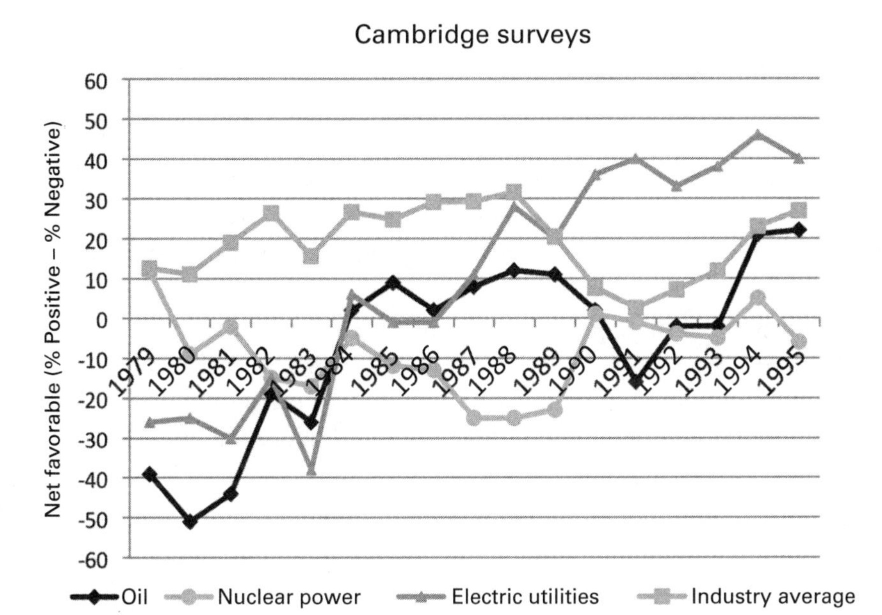

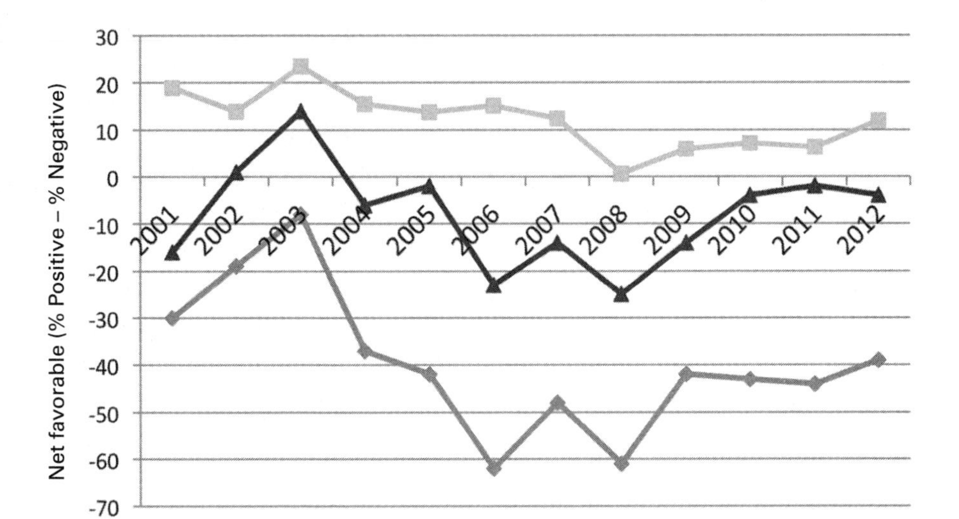

Figure 3.3
Net favorable opinions about energy companies
Sources: Cambridge Reports, Gallup.

favorability, while negative values a net negative favorability. The top panel shows these net favorability ratings from the Cambridge surveys, and the bottom panel the ratings from the Gallup surveys.

Consider the earlier, Cambridge time-series first. Throughout this period, on average, people held net positive views of the industries asked about (alcohol, aluminum, automobile manufacturers, banking, chemical, computer, food, garbage, health care, insurance, lumber, pharmaceutical, plastics, real estate, retailers, telecommunications, and tobacco), though there was a perceptible drop in late 1980s. Regarding perceptions of the three energy-related industries, each saw some large changes over the period covered. The net favorability rating of the oil industry was quite low in the late 1970s and early 1980s, which is not surprising given that this was during the second oil crisis. At its low point in 1980, 74 percent of the public viewed the oil industry very or somewhat unfavorably, compared to just 23 percent who viewed it as somewhat or very favorable. The net favorability of -51 was the lowest for any single industry across all years of the time-series, and these numbers are consistent with Meg Jacobs's argument that many folks believed that oil companies were among those actors chiefly responsible for the oil crisis of the late 1970s. Perceptions of the oil industry did demonstrably improve from about 1985 to 1995, registering many years of net positive values. There was a significant decline in favorability for oil companies in the early 1990s, coinciding with the Exxon Valdez spill in Alaska, but the overall negativity toward the industry was short-lived and not nearly as deep-seated as during the initial years of the Cambridge surveys.

Perceptions of the nuclear power industry were somewhat less volatile than those of the oil industry during the period of the Cambridge surveys, but there were a couple of notable shifts downward in favorability toward nuclear power companies following the accidents at Three Mile Island and Chernobyl. And, unlike feelings toward the oil industry, the net favorability was negative for almost the duration of the period covered in these data.

Views of the electric utilities reflected in the Cambridge surveys showed the most overall change. While perceptions of electricity providers were strongly negative during the first half of the 1980s, there was generally an increasing net positive trend over the period, and from 1987 to 1995, strong (roughly 20–40 percent) positive views of the industry. Of the energy sectors examined in these surveys, the electric utility industry was the only one to surpass the overall average across industries.

The bottom panel of figure 3.3 shows similar net favorability values for 2001 to 2012 for the electric and gas utility and the oil and gas industries, as well as the overall average of the industries covered in Gallup's surveys (accounting, advertising, airline, automobile, banking, computer, education, farming, grocery, healthcare, Internet, legal, movie, pharmaceutical, publishing, real estate, restaurants, retail, sports, telephone, television, and travel.). As was the case in the Cambridge data, on average, the industries covered in the surveys enjoyed net positive ratings. There was a perceptible decline in 2007 and 2008, coinciding with the financial crisis on Wall Street and the overall economic downturn, but overall there were still, on average, positive perceptions of industry.

The same cannot be said regarding perceptions of electric and gas utilities and oil and gas companies. For most of the decade, views of these industries were strongly negative. Interestingly, the trend lines representing the net favorability for these industries are nearly identical, increasing from 2001 to 2003, declining sharply through 2006, and then stabilizing somewhat through 2012. The obvious difference is that views of the oil and gas industries were much more negative. From 2004 to 2012, the average net favorability rating for the oil and gas industry was almost -50, meaning that for every American who viewed these industries positively, three others viewed them negatively.

Collectively, these data suggest a general lack of confidence in the energy and electricity sectors. Energy companies have taken on greater and greater responsibility for electricity delivery, and public utilities have pulled back. This has created a credibility gap for the energy sector, as the public distrusts energy companies, and trusts public utilities far more. One of the great unanswered questions, then, is who the public will trust to lead the development of the energy future.

Energy Preferences versus Energy Realities

To this point, we have examined what energy future Americans would like to see. An important, but very different question is: What

Table 3.3
Expectations of sources of future electricity

Energy source	A lot (More than 25%)	Some (10–25%)	Not much (5–10%)	Very little (Less than 5%)
Coal	37.2%	39.2%	16.6%	7.0%
Natural gas	38.9	49.1	9.9	2.2
Oil	41.0	39.8	14.4	4.7
Nuclear	28.1	39.5	24.8	7.7
Hydro	15.1	43.9	31.4	9.6
Solar	20.0	27.2	32.5	20.3
Wind	19.2	26.6	33.5	20.7

Source: 2008 MIT/Harvard Energy Survey.

energy sources do Americans believe will actually be relied on in the coming years and decades? And do their beliefs coincide with their preferences?

To gauge the public's expectations about what mix of fuels the United States will rely on in the future, the 2007 and 2008 MIT/Harvard Energy Surveys asked the following question:

> Regardless of whether you want more of any particular fuel source, how much do you think the United States will rely on each of the following fuels for electricity over the next ten years?
>
> Coal? Nuclear? Natural gas? Oil? Dams? Solar? Wind?

Table 3.3 presents the results from the 2008 responses (the distributions of responses are very similar for 2007). About four in ten Americans believe that the country will rely a lot (more than 25 percent) on traditional fossil fuels—coal, natural gas, and oil. Another four in ten (five in ten in the case of natural gas) think the United States will rely some (10–25 percent) on these fuels. Regarding nuclear power, almost 80 percent of the public believes that the United States will rely either a lot or some on this energy source for electricity generation. Comparatively fewer Americans see renewable energy sources such as hydroelectricity, solar, and wind power comprising more than 25 percent of future electricity generation capacity. This sense of the American people suggests a clear recognition that these sources, even at full scale, are unlikely to comprise a large amount of total electricity generation, although we cannot say from these data alone whether this reflects an understanding about natural and technological limitations, the politics of energy development, or a stiff dose of realism.

We see a gap between what people want to happen in the future and what they believe is most likely to happen. Americans want to see a shift in the mix of energy sources used to generate electricity. They express overwhelming support for renewable sources of energy, especially wind and solar power, and even increasingly nuclear power. Yet they are also quite skeptical that such a shift will actually take place. When asked what they think the United States will use to power electricity generation, they firmly believe that there will be a continued reliance on conventional fossil fuels. Americans express a clear and strong demand to reduce coal and oil and increase solar and wind and, to a lesser extent, natural gas. Americans, however, do not believe that industry is poised to develop that future in the near term. The slippage between expectations and desires is an unmet demand.

What does that demand reflect? Does it reflect a failure of the energy market to deliver the right product? That is, does the public want a different sort of energy "good" than the market is able to deliver? Or does public demand for more solar and wind and less coal and oil reflect misperceptions on the part of the public about the economics of delivering energy? The answer, interestingly, is some of both, as we learn in the next three chapters.

4

Price and Consequence

Coal, oil, natural gas, nuclear, hydro, wind, solar—those are the alternatives. But that is not really what people want. People want to light and heat their homes, operate computers, televisions, and other appliances, and drive their cars without paying too much. Firms want reliable energy that is not subject to disruptions and will not eat into their profits. People also want to avoid the health effects and safety risks that come with the pollution from some forms of energy. Pollution can be visible and tangible. Residents of Los Angeles closely watch the smog index, especially in late summer, and Angelenos experience unusually high rates of asthma. And, most people would not think of swimming in the rivers around major cities. Visible pollution and health harms have become the rallying cry for government regulation of pollution emissions, especially from power plants, factories, and cars. In short, people want the convenience of a stable and inexpensive energy supply, and they value a clean environment that will not harm their health.

It is our contention that when people say they want more solar and wind or less coal and oil, they are really expressing how they view the attributes associated with those fuels. Why is solar power intuitively desirable? Perhaps the answer is because the sun is nature's source of light and warmth. Why is coal king? Likely, because it is plentiful and has provided an efficient and inexpensive source of power for generations, a time span so long that coal has become virtually synonymous with industry and economic growth.

Our common and intuitive understanding of coal, solar, and other sources of electricity points to a deeper understanding of how people think about energy. People have a sense of what attributes are important to them as well as a perception of these attributes for particular sources of energy. Putting these together, people formulate attitudes and preferences about different energy sources, based on their understanding of these attributes. As an example, let's unpack why most people intuitively dislike coal.

What images immediately come to mind when people think about coal? Perhaps it is the sooty faces of underground coal miners. Or, it might be the large blast furnaces of busy factories making steel and other industrial products. Or, it might be the same factories belching smoke over Pittsburgh or Detroit. Coal is cheap and plentiful in the United States, but mining and burning it to generate electricity is also dirty. Our images of and attitudes toward coal reveal three facts about how people think about energy. First, they think about energy sources in terms of attributes that we value. Second, people assess a source of energy along each of those attributes or dimensions. Coal does well on a cost dimension, being inexpensive and plentiful, but does poorly on an environmental dimension. Third, we weigh these different attributes. How important is cost to an individual? And, how important is air or water pollution?

One might go through the same exercise for oil, nuclear, wind, or solar power. Different images and understandings come to mind about each, but usually the same attributes—prices and environmental consequences. From the perspective of consumers and voters, economic and social costs are the most important attributes, and most people readily have some idea about how well each fuel delivers on each. There are other attributes as well, such as safety, security (i.e., energy independence), and visual aesthetics, but those usually have less weight in consumer and voter decisions.

Survey researchers have long recognized that people want clean and cheap energy. Public opinion polls, however, usually frame energy as a trade-off between energy development and environmental protection. Starting from this assumption that there is an energy-environment trade-off, Gallup and other survey firms that ask these questions track how people would balance these competing claims.

But why the trade-off? Ultimately, people want more of both attributes: we want a cleaner environment and we want cheap energy. The trade-off arises, as we discussed in chapter 2, because of the limitations on current technology, including power plant design, fuels, and delivery systems. Do people see and understand those constraints? Do they think that some fuels are superior in all respects to others, or do they see the technology trade-off?

The MIT/Harvard Energy Surveys were designed to go beyond the usual framing of economy versus environment. Rather than force people to make a choice, we asked them about each fuel separately. How expensive do you think each fuel is? How harmful to the environment? And do you think we should use more or less of each of the fuels? Asking these questions allows us to measure how people see the attributes of the fuels, and to relate their perceptions of the fuels to their preferences about what energy we use and what energy policy the government pursues. This approach allows us to map out the fuels and the energy choice in terms of economic costs and social costs, as we presented them in chapter 2. Here, we will discuss how people perceive the economic costs and environmental harms associated with the various energy sources. In the two chapters that follow, we measure how much value people place on economic costs and environmental costs. How sensitive are they to economic costs? How sensitive are they to the social costs associated with environmental harms? When people do have to face trade-offs, the value or weight that people place on economic and social costs shapes how they make those trade-offs.

Energy Development versus Environmental Protection

We begin with what came before. Previous survey research has examined public concern about the environment and about energy development or the economy generally, rather than in terms of specific energy sources. To the extent that general questions about the economy and the environment can be linked to particular fuels, they can enlighten us about what energy future the public wants. When national surveys have asked questions about specific forms of energy, they typically focus on current events or controversies, including oil spills and nuclear accidents, or strip mining for coal, and drilling for oil and gas in the Arctic National Wildlife Refuge. These types of questions inform our understanding of an immediate crisis or event, but they do not clarify the underlying value that people place on the two key attributes of energy, economic cost and environmental harm.

Eric R. A. N. Smith's work on this subject stands out as the most comprehensive analysis of the traditional survey questions about energy and the environment. Smith compiled hundreds of surveys, mostly conducted by media organizations and polling organizations such as Gallup and Pew, about energy and the environment, and used these data to map out long-term trends in public attitudes about energy use and its environmental consequences. He concludes that people's attitudes are fairly steady; they are in an equilibrium of sorts. Occasional events, such as the grounding and resulting oil spill from the Exxon Valdez tanker in Prince William Sound off the coast of Alaska, can shift opinions about energy versus the environment. Usually those events are ephemeral, and public opinion returns to the steady state it was in before the event. Some rare events lead to more enduring changes in people's attitudes.[1] Accidents at nuclear power plants, such as Three Mile Island or Chernobyl, for example, intensified opposition to the expansion of nuclear power. Such events reveal information to people—such as the safety of nuclear power plants—that they did not fully understand or incorporate in their assessments.

Three strands of survey research map the evolution of public attitudes generally about energy and the environment.

First, and most important, Gallup, Pew, Roper, and other organizations have asked people whether they would prefer less environmental regulation and more energy development or less energy development and more environmental regulation: in other words, Environment versus the Economy. This is a useful starting point for understanding how people view the economic costs and environmental harms associated with energy. That trade-off tells us something about the balance between these concerns, and long-term trends reveal important movement in public perceptions of these two essential attributes of our energy system. As we will see, these questions also produce results that seem to contradict the wide public support for solar and wind power, a puzzle we hope to resolve.

Gallup has asked about the energy-environment trade-off for over a decade. The time-series of Gallup questions is shown in figure 4.1. Until the latter part of the 2000s, close to a majority (if not a majority) of

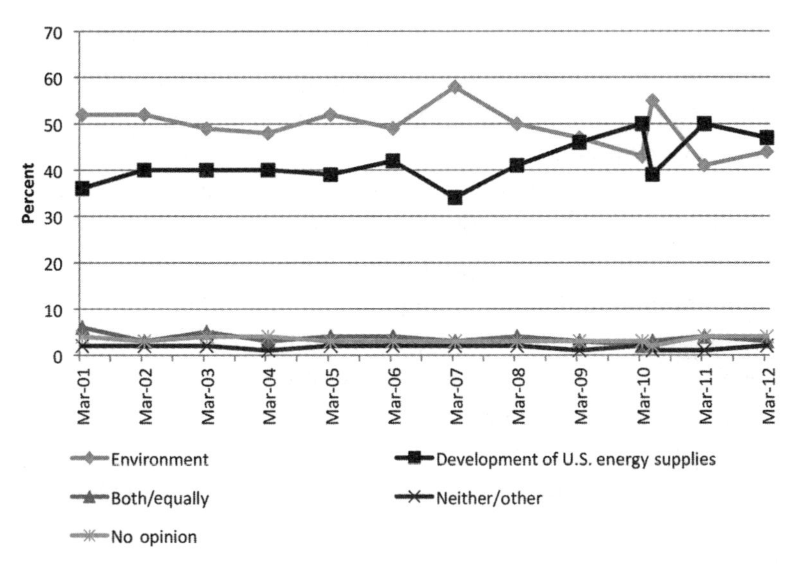

Figure 4.1
Prioritization for environmental protection or energy production. Response to question: "With which of these statements about the environment and energy production do you most agree—protection of the environment should be given priority, even at the risk of limiting the amount of energy supplies—such as oil, gas, and coal—which the United States produces or development of U.S. energy supplies—such as oil, gas, and coal—should be given priority, even if the environment suffers to some extent?
Source: The Gallup organization, various years.

people favored environmental protection over energy development. In March 2009, the gap was within the margin error of the survey, and a year later, for the first time a majority of respondents gave higher priority to energy production than to environmental protection. There was a sharp but ephemeral decline in support for energy development in response to the April 2010 Deepwater Horizon oil spill in the Gulf of Mexico. In response to the spill, Gallup included this question in a May 2010 survey. This is another example of the sensitivity of Americans' opinions on energy issues to focal events such as accidents and oil price shocks, as has been demonstrated in the work of Eric Smith. Opinion then reverted back to the historical levels for one year, before returning to majority support in 2011 and 2012 for development of U.S. energy supplies, despite environmental impacts.

Gallup has also asked a related but somewhat different question over the same period: "Which of the following approaches to solving the nation's energy problems do you think the U.S. should follow right now—emphasize production of more oil, gas, and coal supplies or emphasize more conservation by consumers of existing energy supplies?" The implicit trade-off is a bit different, but fundamentally the question sets energy against the environment. For the entirety of the time-series, Americans have expressed a preference for emphasizing conservation over production. Other surveys going back to the 1970s and 1980s have also found strong support for energy conservation and efficiency measures.[2] The gap between conservation and production has narrowed substantially in recent years, but in the latest survey in the series (March 2013), a majority (51 percent) of the public still indicated a preference for emphasizing conservation.[3]

The Pew Research Center for the People & the Press also began asking about the environment versus the economy in the spring of 2001. Pew asked a slightly different question from Gallup: "Right now, which one of the following do you think should be a more important priority for this country . . . protecting the environment or developing new sources of energy?" The response data are shown in figure 4.2. For the entirety of the Pew time-series, at least a plurality of Americans indicated that developing energy sources should take a higher priority than environmental protection. And, beginning with a survey in September 2005, there was marked increase in support for developing energy sources, an uptick that continued through June 2008. This was a period of extraordinary shocks in oil prices, and it is reasonable to assume that the heightened support for developing energy was tied to the price increase. In the spring of 2005, a barrel of oil cost about $40. Three years later, it had tripled to $120 barrel, eventually reaching a high of $147 in July 2008.[4]

Economics versus the environment is the most common framing of the issues in public opinion research and public debates about energy. Regardless of which survey organization's questions one examines, a consistent pattern emerges. Over the past decade, a noticeable decline in the propensity of the people to choose the environment over energy development has occurred, and during the economic downturn that began in 2008, a majority of the public expressed greater support for

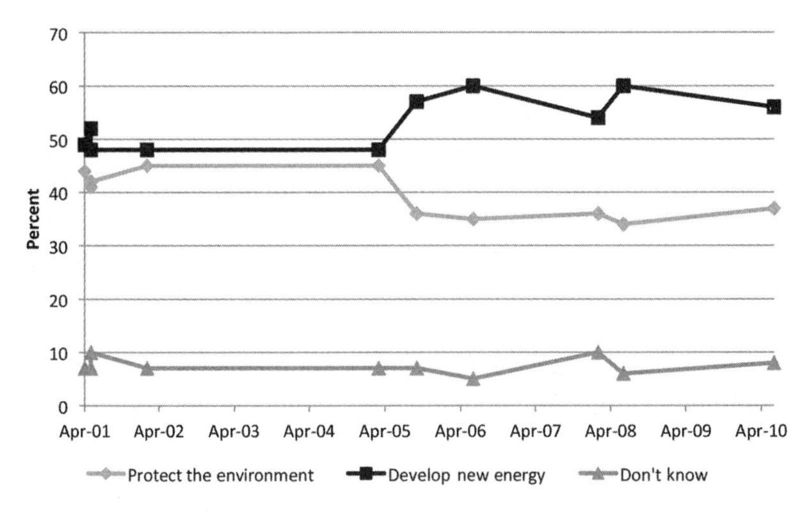

Figure 4.2
Emphasis on energy production or conservation. Response to question: "Which of the following approaches to solving the nation's energy problems do you think the U.S. should follow right now—emphasize production of more oil, gas, and coal supplies or emphasize more conversation by consumers of existing energy supplies?"
Source: Pew Research Center for the People & the Press.

energy development.[5] Today, public preferences about energy versus the environment are split about evenly.

A second key indicator is concern about specific environmental problems. Since the mid-1960s, survey researchers have followed public opinion about various environmental issues. Gallup, Harris, Roper, and other survey organizations have asked people what the most important environmental problem is or how concerned they are about various environmental harms. Since the late 1980s, Gallup, for example, has asked people to indicate how much they personally worry about different environmental problem: "a great deal, a fair amount, only a little, or not at all?" Among the options are pollution of rivers and lakes, air pollution, loss of tropical rain forests, contamination of soil and water by toxic waste, urban sprawl, damage to the Earth's ozone layer, and, more recently, global warming.

Trends in concern about these environmental issues reveal two important lessons. First, throughout the twenty-five-year time-series of this question, clean water has been the public's number one objective, followed by clean air. Second, concern about the environment across all issues has trended downward over the past ten to fifteen years. In the 1980s, over 70 percent said they worried a great deal about clean water; today less than 50 percent do. In May 1989, 63 percent of Gallup respondents reported they were very concerned with air pollution, but by March 2013 that figure had fallen to 36 percent. In May 1989, 41 percent said they were very concerned with ozone depletion, but by June 2008 (the last time this particular environmental problem was included), only 23 percent said they were very concerned this issue. We discuss these data in more detail in chapter 7, especially as they relate to worries about global climate change.

A third, and final, indicator of note is what is known as the MIP or most important problem. "What is the most important problem facing the nation today?" In seventy years of surveys conducted by the Gallup, the economy and jobs almost always top the list. In July 2013, 52 percent offer some form of economic issue as their top concern. Excluding the deficit and other matters, 42 percent say the economy generally or unemployment and jobs specifically. Energy and the environment, on the other hand, barely register. Only 1 percent volunteer that gas prices are the most important problem facing the nation, and just 1 percent indicate the environment or pollution. The MIP is often criticized for reflecting "top of the head" responses, for not measuring more deeply held preferences, or for being susceptible to framing effects.[6] However, even when we use closed-ended questions, the economy appears to trump the environment. The MIT/Harvard Energy Surveys have regularly offered respondents a list of twenty different issues, ranging from unemployment and economic growth to crime and war to abortion and the environment. The environment routinely ranks around twelfth to fourteenth. The economy has been the number one or number two issue throughout the past decade.

The indicators of public opinion about energy and the environment are clearly consistent with Smith's assessment. There is a steady state of sorts that is moved by short-term events, such as gas price spikes, recessions, and accidents. Underlying that steady state, we argue, are two attributes—perceptions of the economic and social costs of energy. Americans have long been concerned about the environment, often greatly concerned, but that issue is not always a top-tier problem. General concerns about the economy usually take center stage. Even within the area of energy, price and environmental harm seem to compete with each other. Over the past two decades there has been declining concern about the environment, and today economic and environmental concerns appear to weigh equally in people's thinking.

But these questions take us only so far in understanding how Americans think about our energy future. They reveal that people are concerned about both the economic and environmental consequences of energy use. They do not, however, tell us whether people have any idea about the economic or environmental consequences of particular energy sources, or whether people's beliefs about the economic and social costs of using coal, natural gas, wind, oil, solar, or nuclear power affect their attitudes toward those fuels. We also do not know what these beliefs mean for the energy choices we face. Does the rise of economic concerns spell a resurgence of coal? Does it mean the demise of public support for wind and solar power?

Indeed, there seems to be a puzzling inconsistency between the energy versus environment survey research and our findings in chapter 3. The American public has become more attentive to the economic costs associated with energy and less attentive to environmental harm over the past decade. Yet Americans strongly embrace relatively expensive solar and wind power and increasingly view with disfavor much cheaper coal and oil. Declining support for coal appears incompatible with rising concern about economic costs of energy and declining worries about the environment. Do people really value the environment less today than they did ten years ago? Are people increasingly motivated by economic considerations? Clearly we need to dig deeper into public attitudes about economic costs and environmental harms in order to get a clearer explanation of how people think about the energy choices.

Perceived Economic Costs

Americans are acutely aware of energy prices. The average American does not know the national unemployment or inflation rates, but most know the local gas price within a few cents and last month's electricity bill within a few dollars.[7] Americans' heightened awareness about energy prices owes partly to regular exposure. We see gas prices many times a day, driving to work or to the store. We see (and presumably pay) electricity bills each month.

But how aware are people of the cost of generating electricity from particular sources of power? There is no reason to expect that people would have any sense of the cost of providing electricity from various fuels. Many energy companies have started to offer consumers the option of buying electricity from wind or solar installations, at a higher price. And if you have looked into putting solar panels on your roof, you might have a good sense of the cost of that power source. Otherwise, people do not make direct decisions about what power they use. Most people we casually talk with about energy in our hometowns (Cambridge, Massachusetts, and Washington, D.C., respectively) do not know where the closest power plant is and what sort of fuel it burns. Even still, if people are to make sensible decisions about energy, some basic knowledge or at least reasonably accurate beliefs would seem essential. Do people have any idea what the cost of electricity from particular fuels might be? Does the public, on the whole, get the prices right?

The MIT/Harvard Energy Surveys asked people to evaluate each of seven energy sources—coal, natural gas, oil, nuclear, hydro, solar, and wind—according to its affordability and environmental cleanliness. First, regarding people's perceptions of cost, we asked the following question:

How expensive do you think it is to produce electricity from each of the following fuels?
Coal, Natural gas, Nuclear power, Oil, Dams, Solar, and Wind

Very expensive
Somewhat expensive
Moderately priced
Somewhat cheap
Very cheap

The full distribution of responses in the surveys for each type of energy is displayed in table 4.1.

Most Americans hold beliefs about what the cost of electricity would be if it were provided by one of the fuels in question, and people see big differences in cost across the fuels. In general, perceptions of the cost of electricity from each power source do not differ markedly across the four survey years, though there is some variation, roughly in line with fluctuations in electricity costs. The public is not certain about electricity prices associated with fuels. The 2002 survey did not include a "Not sure" response option, but the later years did. The percentage of people saying they were not sure of the cost of electricity from a given fuel ranged from 15 to 22 percent, depending on the energy source. This rate of "Not sure" responses is entirely reasonable for public opinion polls.

The MIT/Harvard Energy Surveys reveal a clear ordering of fuels in terms of perceived costs. There are three groupings of fuels—those seen as most expensive, those seen as involving moderate cost, and those seen as cheap.

Unmistakable is the fact that people view oil and nuclear power as the most expensive. The perceptions of oil are easy to understand, given that the price of oil has been at historically high levels for much of the last decade. Oil is also the type of energy to which Americans pay the most attention, given the coverage it gets by the news media and the weekly visits most people make to the gas pump to fill their cars and trucks. The case of nuclear power is less obvious. In the last three surveys, a near majority of the respondents indicated that they believed nuclear power was either very or somewhat expensive. (The total was 71 percent in 2002 when "Not sure" was not included among the response options.) Few Americans, in the range of 10–20 percent, view nuclear power as inexpensive. One might speculate that the public views nuclear as expensive because it is likely the way of generating power that it understands the least. It is also possible that people think about the expense of nuclear weapons or the cost of building nuclear power plants. We cannot say for sure based on these data, but regardless the perception that nuclear power is very costly is widely held.

Coal, natural gas, and to some degree hydropower comprise a middle category. With respect to coal, the modal response in the first three surveys was moderately priced, with the rest of the responses roughly evenly split between expensive and cheap. Coal was perceived to be considerably less expensive in the 2011 survey. Perceptions of the costs of natural gas are similar in that the most frequent response is moderately priced, but the distribution of responses is skewed more toward the expensive side of the ledger, until the 2011 survey when more individuals viewed natural gas as less expensive, reflecting perhaps the severe drop in the price of natural gas resulting from the enormous growth in development of shale gas.

Table 4.1
Perceived economic cost of energy sources

2002 sample

	Very expensive	Somewhat expensive	Moderately priced	Somewhat cheap	Very cheap	Not sure
Coal	13.4%	24.5%	35.1%	21.4%	5.6%	–
Natural gas	11.8	32.8	42.5	11.5	1.4	–
Oil	25.2	42.1	26.7	5.3	0.7	–
Nuclear	38.4	33.0	19.3	7.4	2.0	–
Hydro	9.9	24.1	34.7	22.4	8.9	–
Solar	9.9	19.4	22.7	28.1	19.9	–
Wind	4.5	11.6	19.3	31.1	33.5	–

2007 sample

	Very expensive	Somewhat expensive	Moderately priced	Somewhat cheap	Very cheap	Not sure
Coal	11.1%	18.1%	26.1%	18.5%	8.8%	17.6%
Natural gas	15.4	28.4	29.1	10.5	1.3	15.3
Oil	28.7	30.9	19.1	5.5	1.2	14.7
Nuclear	26.7	23.4	15.9	8.9	4.6	20.5
Hydro	5.6	13.9	30.4	21.2	9.9	19.1
Solar	9.3	16.3	17.1	20.0	22.1	15.1
Wind	5.3	12.7	16.8	20.4	27.9	16.9

2008 sample

	Very expensive	Somewhat expensive	Moderately priced	Somewhat cheap	Very cheap	Not sure
Coal	10.6%	16.8%	28.8%	18.2%	5.1%	20.5%
Natural gas	13.5	25.0	32.1	10.0	1.3	18.1
Oil	29.7	31.5	16.4	4.0	1.1	17.2
Nuclear	23.7	22.7	15.8	11.9	4.1	21.8
Hydro	6.3	12.3	28.0	22.0	9.2	22.2
Solar	7.0	12.8	20.5	24.3	19.7	15.8
Wind	4.3	9.9	19.6	23.9	24.2	17.9

Table 4.1 (continued)

2011 sample

	Very expensive	Somewhat expensive	Moderately priced	Somewhat cheap	Very cheap	Not sure
Coal	7.7%	11.7%	24.1%	24.3%	13.8%	18.4%
Natural gas	6.9	15.5	29.1	23.4	7.7	16.9
Oil	19.2	24.9	24.3	11.5	3.6	16.1
Nuclear	18.4	21.2	20.6	12.0	8.7	19.2
Hydro	4.5	10.0	26.9	25.5	13.7	19.5
Solar	14.4	16.5	15.0	17.6	21.4	15.0
Wind	11.7	17.2	14.3	17.4	23.7	15.7

Source: MIT/Harvard Energy Surveys.

Surprisingly, solar and wind power are viewed as the least expensive. In fact, the average rating given to these power sources by Americans is that they are "Somewhat cheap." In the 2007, 2008, and 2011 surveys, just under 50 percent of the respondents characterized generating electricity from these renewable sources as either somewhat or very cheap. In the 2002 survey (without the "Not sure" option), the totals were even higher, reaching nearly 65 percent for wind. Only about 20 percent of the respondents in the surveys indicated that solar and wind was either very or somewhat expensive, although this rose to about 30 percent in the 2011 sample for solar power. Although the costs of generating electricity from wind have declined over the last decade, it remains at best on par with conventional fossil fuels (without taking social costs into account), while solar is considerably more expensive.

Another way to compare perceptions of cost across sources is to consider the average response across the survey respondents. Figure 4.3 displays the mean value of the responses across the samples, where responses of "Very cheap" are set to 5, "Somewhat cheap" to 4, "Moderately priced" to 3, "Somewhat expensive" to 2, and "Very expensive" to 1 ("Not sure" responses are excluded). The higher the calculated value, the *less expensive* is the perception of the costs of generating electricity for that source. Across the surveys, the rank orderings of these average values is almost identical. Oil was considered be to the most expensive way to generate electricity, except in the 2002 sample, when nuclear power was

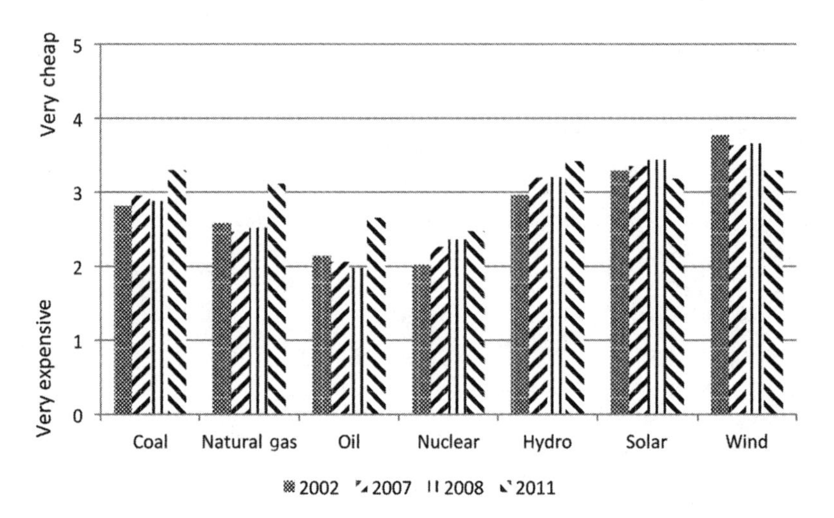

Figure 4.3
Average perception of cost
Source: MIT/Harvard Energy Surveys.

perceived to be more expensive. Following these two sources, people viewed natural gas as the next most expensive, followed in order by coal, hydro, solar, and wind. This graphic reinforces the key finding that people believe that renewable technologies are the least costly ways to produce electric power.

How do these perceptions of costs line up with actual costs? The answer of course depends on which type of costs you consider. When answering this question, we believe people are thinking about the cost of electricity from an already existing power plant, rather than the costs of extracting the fuel source, constructing the plant, or delivering the electricity to end users, and so forth. Moreover, we think that people are making assessments based on their views of the costs as reflected in the market, and not the costs of any externalities that they think are associated with using a particular fuel. We return to this subject in more detail in chapter 6. With this in mind, a few interesting patterns emerge.

First, the public provides the right rank order of the costs of the fuels most commonly used to generate electricity. Americans tend to view coal and natural gas as much less expensive than oil and nuclear power. One can see in the data a precipitous drop in the perceived cost of natural gas from 2008 to 2011, corresponding to a true drop in the price of these fuels.

Second, respondents to our surveys state the cost of nuclear power is higher than that of other conventional energy sources. When we have presented these facts to our colleagues in nuclear engineering, we get an argument about what is reality. Although nuclear plants are more expensive to construct, once they are in operation, they produce electricity at a much lower marginal cost than either natural gas or coal, yet respondents indicated a strong belief that nuclear power is more expensive. From the consumer's (and survey respondent's) standpoint, the price of electricity is the right criterion. The electricity price associated with nuclear power is usually higher than for coal and natural gas, because nuclear power must recoup the much higher cost of financing the construction of the power plant itself.

Third, and most stunning of all, the public holds completely wrong beliefs about the costs of solar and wind power. A very large majority of the American public believes that solar and wind power is "Very cheap" or "Somewhat cheap," and they think that these sources of energy are much less expensive than coal, natural gas, nuclear power, and oil. If the public were right, the nation would quickly wean itself from fossil fuels. The degree of misperception is particularly stunning for solar power, which, on a utility scale, remains by far the most expensive way to generate electricity, at perhaps five times the cost of coal. It seems that people make the assumption that, because the cost of the input is free, it must be a cheap way to produce power.

One possible explanation is that people are projecting their dislike of coal, oil, or nuclear power onto solar and wind. That is, they want solar and wind to be cheap because they see them as the opposite of fossil fuels, and they project (in the psychological sense of that word) good attributes onto those fuels out of antipathy to the fuels they want the nation to get away from. We will show later in this chapter that beliefs about coal, oil, and nuclear power are completely uncorrelated with beliefs about solar and wind, so such projection cannot take place. The misperception seems to reflect a lack of familiarity or understanding, and most likely a bad case of wishful thinking.

In the next two chapters, we explore how that misperception shapes public attitudes about the willingness to expand the use of energy sources.

Importantly, as people learn more about the price of solar and wind power, they like it less. This fact ought to stand as a caution to environmental advocates who think that if we just informed the public they would follow the movement. Information may, in fact, work against the environmentalist impulse. Informing people of the true cost of coal, oil, natural gas, and nuclear power will likely change aggregate public opinion little because people already seem to know the relative price of these fuels. However, informing people that wind and solar operated at a large scale would lead to electricity prices that are two to five times higher than current prices would be a real jolt to public opinion precisely because people think these fuels are very cheap when they are not.

Perceived Environmental Harms

Environmental harms and their social costs are the second main attribute of energy production and consumption. The 2007, 2008, and 2011 surveys asked about perceived environmental harms in the following way:

Some ways of generating electricity may be harmful to the environment we live in.

How harmful do you think each of these power sources is?

Very harmful
Moderately harmful
Somewhat harmful
Slightly harmful
Not harmful at all

The 2002 survey asked a slightly different question, mentioning air pollution, water pollution, and toxic wastes as particular types of environmental harm. Although we think the question is qualitatively similar to the one in the later surveys, more explicitly noting the types of environmental harms does provide some additional information, since we can say with a little more certainty that the environmental impacts that the respondent is assessing are local, and not global in nature. This is the question asked on the 2002 survey (with the same response categories):

Some ways of generating electricity may be harmful to the environment we live in because they produce air pollution, water pollution, or toxic wastes. How harmful do you think each of these power sources is?

Table 4.2 presents the distribution of responses for each fuel source in the four surveys.

As with perceived costs, the public sees a clear difference between traditional and alternative fuels. Electricity generation from coal is strongly perceived as harmful to the environment, as is nuclear power and oil. A majority of respondents in the 2007 and 2008 surveys viewed coal as very or moderately harmful, and almost two-thirds of the sample had this perception in 2002 (without a "Not sure" response). Just as striking, less than 15 percent of the respondents in each survey characterized coal as being either slightly harmful or not harmful at all to the environment. There was a bit of softening of concern in the 2011 survey, with just over 40 percent of the public characterizing coal as either very or moderately harmful. The perceived harms of oil and nuclear power as electricity generation sources are generally similar to those of coal, and nuclear power was perceived as the most harmful in 2002. We discuss perceptions about nuclear power more later in this chapter, but it is important to note that the 2011 survey was administered about six months after the Fukushima Daiichi accident in Japan, and perceptions of the harms from generating electricity from nuclear power actually *declined*. Natural gas sits alone in a middle category, being viewed on average as "Somewhat harmful" or "Slightly harmful."

The public's view about renewable sources is much more favorable. With respect to solar and wind power, at least 70 percent of the public in each of the surveys characterized these energy sources as not being harmful at all to the environment. Hydropower is close behind, with upward of 60 percent of the respondents indicating that using it to produce electricity was either only slightly harmful to the environment, or not harmful at all.

Figure 4.4 displays the average perceptions of environmental harms for each of the sources across the surveys, using the same methodology that we did above for costs (in this case, higher values indicate a perception that a fuel source is *less harmful*). These distributions suggest a clear and relatively stable rank ordering of perceived harms. The public clearly and rightly views the environmental impacts of renewable technologies as minimal. The average values for these fuels indicate that the public views hydropower as "Slightly harmful" and solar and wind as "Not harmful at all." Coal, nuclear, and oil are at the other end of the

Table 4.2
Perceived environmental harm of energy sources

2002 sample

	Very harmful	Moderately harmful	Somewhat harmful	Slightly harmful	Not harmful	Not sure
Coal	32.9%	31.7%	24.2%	9.0%	2.3%	–
Natural gas	6.9	18.0	35.0	29.4	10.8	–
Oil	23.4	37.1	28.0	8.6	2.8	–
Nuclear	45.1	22.5	17.3	10.4	4.7	–
Hydro	6.0	12.0	19.0	29.2	33.8	–
Solar	2.7	3.1	8.9	14.0	71.2	–
Wind	1.7	2.9	6.9	12.8	75.8	–

2007 sample

	Very harmful	Moderately harmful	Somewhat harmful	Slightly harmful	Not harmful	Not sure
Coal	29.4%	25.4%	22.5%	8.3%	5.2%	9.2%
Natural gas	4.7	15.8	30.6	23.6	14.7	10.7
Oil	24.2	26.2	25.4	12.2	4.5	7.6
Nuclear	36.5	14.3	15.7	14.0	9.8	9.7
Hydro	2.7	6.5	15.7	24.1	40.2	10.7
Solar	1.9	1.8	4.4	8.4	75.5	8.1
Wind	2.2	1.8	5.5	10.1	72.6	7.7

2008 sample

	Very harmful	Moderately harmful	Somewhat harmful	Slightly harmful	Not harmful	Not sure
Coal	31.7%	22.6%	21.4%	9.9%	3.6%	10.7%
Natural gas	3.3	12.4	27.5	27.4	16.8	12.8
Oil	19.5	25.0	28.4	11.6	5.3	10.2
Nuclear	30.1	16.5	14.1	15.1	12.1	12.2
Hydro	3.9	4.7	15.1	22.7	38.7	14.9
Solar	1.1	2.7	3.7	6.6	78.0	8.0
Wind	1.2	1.0	4.3	9.5	74.3	9.7

Table 4.2 (continued)

2011 sample

	Very harmful	Moderately harmful	Somewhat harmful	Slightly harmful	Not harmful	Not sure
Coal	24.2%	18.8%	20.3%	15.8%	8.3%	12.6%
Natural gas	3.1	10.7	18.3	25.4	30.0	12.5
Oil	15.3	22.9	20.9	19.7	9.8	11.4
Nuclear	27.1	15.2	14.3	17.0	14.5	12.0
Hydro	3.1	6.5	14.5	22.4	38.7	14.8
Solar	1.5	1.5	3.6	9.2	74.2	10.0
Wind	1.1	1.8	4.3	12.8	69.9	10.1

Source: MIT/Harvard Energy Surveys.

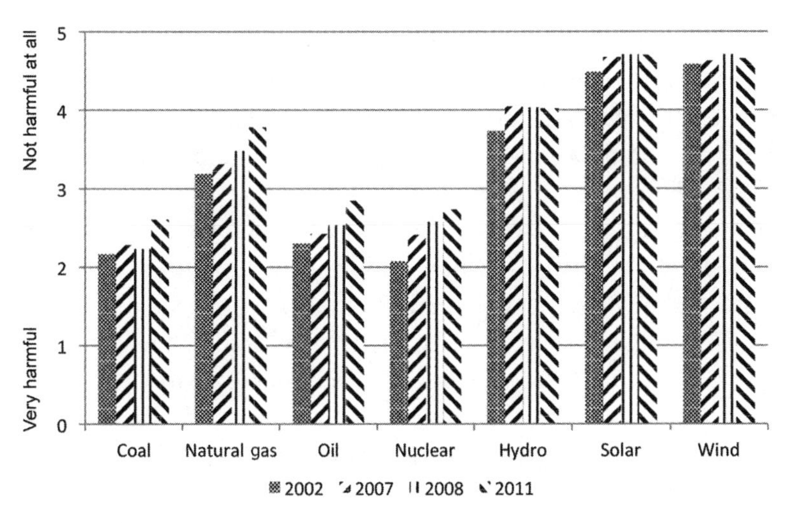

Figure 4.4
Average perception of harm
Source: MIT/Harvard Energy Surveys.

spectrum; all are viewed on average as being between "Moderately harmful" and "Somewhat harmful." And, as was the case with perceived costs, natural gas is right in the middle, with the average belief being that this fuel source is between "Somewhat harmful" and "Slightly harmful" to the environment.

It is also interesting to note that these beliefs do not seem to have changed much over time. One piece of evidence to this effect comes from a survey conducted by Cambridge Research in 1990 that asked respondents to evaluate the environmental threat on a scale from 1 (no threat) to 7 (large threat) stemming from different energy sources used to generate electricity. The results (i.e., average for full sample), in order of most harmful, were nuclear power (5.2), coal (4.9), oil (4.8), natural gas (3.8), hydropower (3.0), and solar (2.5),[8] which is very similar to what we see today.

Beliefs about the environmental harms of different energy sources tend to match reality, though nuclear power seems to evoke an especially strong and perhaps exaggerated public reaction. People tend to correctly perceive coal and oil as having the most adverse environmental effects, and the combustion of these fuels for electricity does have more severe health effects than other sources. Natural gas, by comparison, is seen as posing fewer risks than coal, but more than hydropower and other renewable technologies. And renewable sources of electricity generation, particularly wind and solar, are correctly perceived to pose few environmental harms.[9] Thus, unlike the misperceptions of cost, particularly with regard to renewable fuels, Americans demonstrate a solid understanding of the relative environmental consequences from using different fuels.

The Case of Nuclear Power

Nuclear power is perhaps the most intriguing of all of the modes of electricity generation. The industry in the United States has had a rough ride since the Three Mile Island accident, and the prospect of a nuclear plant being located within twenty-five miles of person's home, even in the same state, provokes strong NIMBY reactions. For those worried about air pollution and global warming, though, nuclear power also holds the prospect of a power source that can operate at the same industrial scale as coal or natural gas, but one that has an almost carbon-free fuel cycle.

Some policy experts concerned about global warming often tout nuclear as an essential part of any solution to the problem.[10]

The public, for its part, is divided over the development of nuclear power. The average American views nuclear power as both very expensive and very harmful to the environment. It has the largest fraction of people choosing the option "Do not use at all" when answering the survey, but a substantial fraction of Americans also support expansion of this fuel. Proponents of nuclear power say Americans have it all "wrong." Producing electricity from nuclear power is inexpensive on a per unit of output basis compared to all other sources of energy. Moreover, the environmental harms typically associated with energy use—conventional air and water pollutants, carbon emissions—are pretty much nonexistent for nuclear power, but people tend to view this form of electricity generation as harmful.

Why?

The question of costs is somewhat debatable, but the question of air and water pollution is not. Nuclear power's cost of electricity include not only the marginal cost or per unit cost of producing a unit of electricity but also the remaining cost of the construction of the power plant, especially interest payments from financing such a large complicated piece of machinery. Environmental harms are a different story, and they reflect other unresolved problems with this technology.

To further dissect people's concerns about nuclear power, several of the MIT/Harvard Energy Surveys asked additional questions about the environmental harms associated with nuclear power. Recall from chapter 2 that one of the features of generating electricity from nuclear fission that separates it from other energy sources is the waste. Safe storage of spent fuel has been a controversial issue in the United States (and, in many other countries) since the onset of commercial nuclear power in the late 1950s. Waste storage poses a particularly thorny problem because some of the most toxic by-products remain a threat to human health and the environment for hundreds or even thousands of years. The United States has not pursued reprocessing as aggressively as some other countries and has instead focused on a strategy of underground storage. Locating an appropriate site and constructing a suitable long-term waste storage facility has proved exceedingly difficult. The waste repository at Yucca Mountain, Nevada, has been largely constructed, but political opposition and geological problems mean that the repository has not, and may never, become operational.

In the minds of many Americans, waste storage is a showstopper for nuclear power. In the 2007 survey, only 25 percent agreed with the statement that "nuclear waste can be stored safely for many years." This was down about 10 percentage points from when we asked the same question in 2002. About two-thirds of the sample indicated that they would support a significant expansion of nuclear power "if there were a safe and effective way to deal with nuclear waste," which suggests that general support for nuclear power would be considerably higher were Americans to have more confidence that the long-term storage problems could be resolved.

The 2011 survey also included a question asking respondents to evaluate the safety of generating electricity from different types of power plants. More than a third (35 percent) of those asked indicate that they think that nuclear plants are "Very unsafe," which is double the percentage saying the same about coal (17 percent) and oil (14 percent) plants, five times that of natural gas plants (7 percent), and ten times that of hydro (4 percent), wind (4 percent), and solar (3 percent) power facilities. Renewable energy facilities, by contrast, are judged by about 70 percent of Americans to be "Very safe," and fossil fuel plants are generally considered to be "Somewhat safe."

Americans appear more concerned about the prospects of extreme events associated with nuclear power plants, such as the catastrophic accidents at Three Mile Island (United States) in 1979, Chernobyl (Russia) in 1986, or Fukushima (Japan) in 2011. In response to the Fukushima accident, Germany and Switzerland decided to begin a process of transitioning away from nuclear power completely. Japan has considered a full-fledged phaseout of nuclear power, although the national government has not yet reached a final decision. And, the United States is undergoing an extensive review of the threat of catastrophic events or accidents to its nuclear fleet.

The public senses a high level of risk of an accident. The 2007 survey asked respondents to assess the likelihood of a serious accident at a nuclear power plant in the next ten years. One in ten Americans responded that they thought it "almost certainly will happen." Another 16 percent thought it was "Very likely," while 28 percent thought it was "Somewhat likely." About 30 percent of the public thought it was somewhat or very unlikely, or almost certain not to happen. The 2002 survey asked the same question with a similar distribution of responses.

Waste, accidents, and other risks are part of the overall environmental and social costs associated with use specific fuels. For most of the fuels we use to generate power, people view the risks associated with waste and accidents as unimportant or as likely to result in air and water contaminants. But waste and accidents are a unique and crippling liability for the nuclear industry in the eyes of most Americans.

One Public?

So far we have treated the public as a whole, as an undifferentiated pool of consumers or voters. We recognize that there is, of course, variation among people in their understanding or beliefs about various fuels. Some people think coal is expensive; others think it is cheap. Some people think coal is dirty; others think it is clean. We have treated these differences among people as unimportant, except insofar as some people think a fuel is expensive or cheap, or harmful or clean. In other words, the differences in beliefs largely reflect true differences in understanding rather than some other factor, such as political party, ideology, or income. That's a big assumption and one for which we expect some disagreement.

Anthony Leiserowitz of Yale's School of Forestry and Environmental Studies has developed a typology of people according to their demographic and political characteristics and their concern about climate change. He argues that there are substantial differences among people, and that those types of people are identifiable on the basis of demographic and political factors, such as age, education, income, gender, political party, and political ideology. In chapter 7, we revisit this question as it pertains to climate change. To what extent do such factors explain variation in people's beliefs about the economic and environmental costs of energy?

The short answer is not much. Factors such as age and political party do register statistically significant correlations with perceptions of costs and harms, but by far the largest factor explaining people's beliefs about the attributes of energy is the type of fuel used. We performed multiple regression analysis in which we predicted perceived harms and perceived costs of each fuel with a large number of demographic characteristics and with fuel types. All of these factors combined explained roughly 40 percent of the total variation in perceived harms and about 20 percent in the costs of specific fuels across all respondents in the 2002, 2007, 2008, and 2011 MIT/Harvard Energy Surveys. The rest of the variation is not explained, and reflects respondents' uncertainty or guessing.

Figure 4.5 displays the percentage of the explained variation in perceptions of harms and costs. The type of fuel accounts for about 90 percent of the variation in perceived harms and 86 percent of the variation in perceived cost. Demographic and political factors account for the

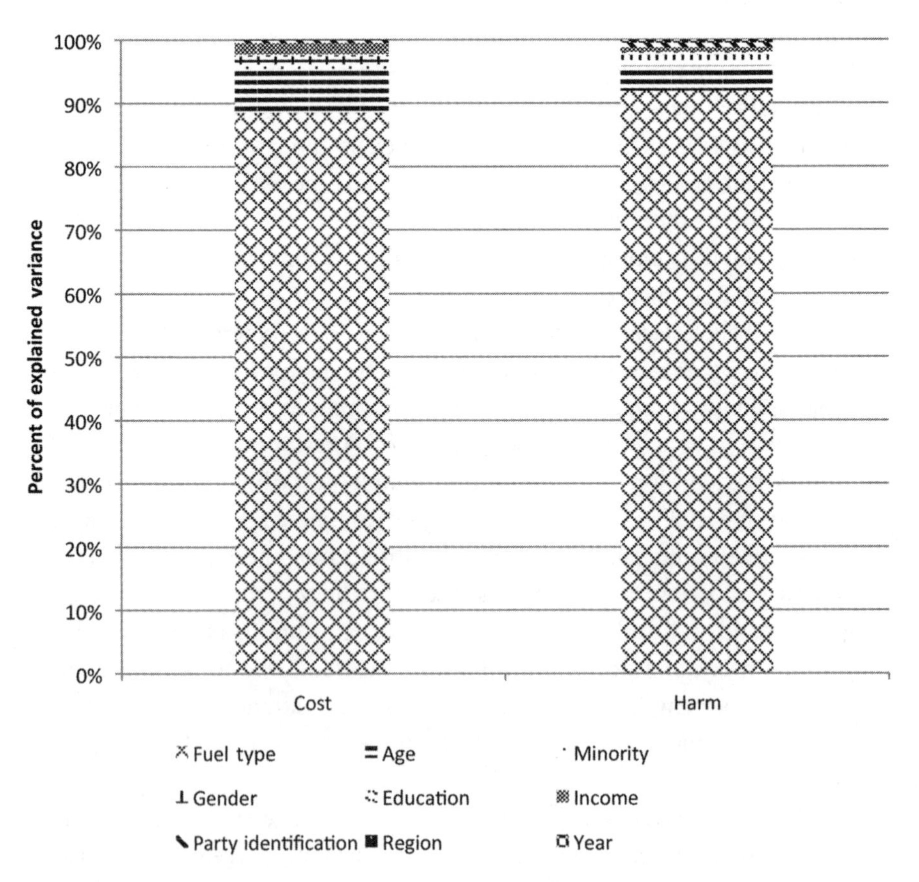

Figure 4.5
Explained variation in perceived costs and harms of energy, across fuels
Note: Results from ANOVA, pooling the 2002, 2007, 2008, and 2011 MIT/Harvard Energy Surveys.

small balance, with age the most important factor, accounting for about 4 percent of the explained variation in perceived harms and 7 percent of the explained variation in perceived costs. In other words, when it comes to explaining people's understanding of energy, the type of fuel is by far the most important factor. The demographic or attitudinal features of a person are of secondary importance, at best.

Of the demographic and attitudinal factors, what matters most? On questions of perceived costs, the main differences arise with age, gender, income, and party identification. (Regression results are presented in tables A4.1 and A4.2 in the appendix). Middle-age respondents (those between thirty and sixty-five) give the highest cost ratings to fossil fuels. For nuclear, hydroelectricity, solar, and wind power, younger people think the fuels are more expensive than older cohorts. Women view solar and wind as less expensive than men do, but perceive the rest of the fuels as more expensive than men do. Democrats tend to view natural gas, oil, nuclear, and hydropower as more expensive, while Republicans on average see these fuels, as well as coal, as less expensive. Again, while these differences are real (in a statistical sense), they are small in magnitude. The difference between young and old, women and men, and Democrats and Republicans represent at most a third of a point on a five-point scale, and in many cases is much smaller. By comparison, the difference between wind and nuclear power is 2.5 points—fully half the entire scale for costs and for harms.

On assessments of harms, we see much larger partisan differences. Republicans tend to say that coal, oil, gas, nuclear, and hydroelectric power are somewhat less harmful than Democrats do, but there are only trivial partisan differences on solar and wind power. The political divisions on coal, oil, natural gas, and nuclear power, however, amount to a difference of half a point on a five-point scale from "Not harmful at all" to "Very harmful." It is important to emphasize, however, Republicans and Democrats are on the same side of these issues. Most Republicans and most Democrats say coal is "Very harmful" or "Somewhat harmful." Most Republicans and most Democrats say that solar is not very or not harmful at all. Age is also a highly important factor, and for most of the fuels the difference between young and old is as large as the difference between Democrats and Republicans. For nuclear power and natural gas, the difference in perceived harm among people under age thirty is about one- to two-tenths of a point on the five-point scale. Again, this is a statistically detectable difference, and one of the larger differences across demographic groups, but it is clearly secondary to the differences across types of fuels. There are also some differences in terms of education and income, but these too are small in comparison to the types of fuel.

The real explanation of public perceptions, then, is not demographic or political differences among people, but the type of fuel in question. Conjectures that significant differences exist across demographic and political groups certainly have some merit. Such factors register statistically significant correlations with perceived harms and costs of various energy sources. However, demographic and political differences are dwarfed by the differences in perceived costs across fuels for all people. Once political debate has settled into a question about coal or nuclear power or alternative energy, the demographics may have some play, but by that time, the particular fuel has already defined the terms of the political debate. Politics and demography do not explain the large differences in people's perceptions of the costs and harms of fuels. Such characteristics of individuals do not explain why the American public substantially underestimates the costs of solar and wind power, or why the American public holds such a negative view of nuclear power. The root explanation for these differences lies not in politics or sociology of the American public, but in the peculiarities of specific fuels.

In this regard, one pattern in our survey data deserves emphasis. In analyzing the data from the 2002 survey, we saw a theme emerge—really a clustering or segmentation in the way people think. Looking across the analyses of the factors that predict perceptions of fuels, we saw that the certain variables correlated the same way with opinions about coal, natural gas, oil, and nuclear power, but that they correlated differently with solar and wind power. The same pattern occurs in every one our surveys.

People think of solar and wind as a type of fuel, call these the *alternative fuels*, and they think of coal, natural gas, oil, and nuclear, and to some extent hydroelectricity, as another type of energy, call these *traditional fuels*. Attitudes, opinions, and perceptions of alternative fuels are correlated with one another, and attitudes, opinions, and perceptions of traditional fuels are correlated with each other. But, and this is the surprising part, attitudes about alternative fuels are *not* correlated with attitudes about traditional fuels.

For example, there is an exceedingly high correlation between people's beliefs about the costs of solar power and the costs of wind power and also between the harms of solar power and the harms of wind power. The correlation among perceptions of costs of these two fuels is 0.9 in the MIT/Harvard Energy Surveys, and the correlation among perceived harms is 0.8. People who think solar is not harmful (and most do) also think wind is not harmful. A correlation of 1.0 would mean that the two factors were identical; a correlation of 0 would mean they were unrelated; and a correlation of -1.0 would mean that the two factors are perfect predictors of each other, but are opposites of each other. (These correlations using the 2008 survey are presented in table 4A.3 in the appendix to this chapter for the curious reader.)

A similar, but less strong, set of correlations binds together perceptions of fossil fuels. Perceived costs of coal, oil, and natural gas are correlated at about 0.5 or 0.6, and the perceived cost of nuclear and hydropower has a correlation with these fuels of about 0.4 to 0.5. People who think coal is cheap also tend to think natural gas, oil, nuclear and hydro are cheap, but the relationship is not as strong as among the alternative energies. Perceived environmental harms of coal, oil, and natural gas show similar correlations to those for perceived costs. The perceived harms of nuclear and hydropower are somewhat less related, with correlations with the other fuels between 0.3 and 0.4.

The correlations between traditional fuels and alternative fuels, surprisingly, are effectively 0. For perceived costs, the correlations between each of the traditional fuels and each of the alternative fuels are statistically indistinguishable from 0, except for a slight positive correlation between hydroelectricity and solar or wind. For perceived harms, the correlation between each of the traditional fuels and each of the alternative fuels is statistically indistinguishable from 0, except for a slight positive correlation between natural gas and solar and between hydroelectricity and solar or wind. In other words, how someone understands the attributes of the traditional fuels is unrelated to how they understand the alternative fuels.

What explains these results? Our intuition is that they derive from different levels of familiarity that people have with energy sources. Most people are well acquainted with the set of traditional fuels used to generate electricity. Over the past fifty years or so, the United States has consistently relied on coal, natural gas, and nuclear power for most of its electricity. Wind and solar were until recently aspirational fuel choices for large-scale power generation; many people were certainly aware of them as future options, but utility-scale development of these sources is a relatively new phenomenon. These differences in familiarity may result in people thinking differently about their costs and harms than they do for coal, natural gas, oil, and nuclear power.

The common misperception that solar and wind are cheap, then, seems to be a feature unique to those fuels, rather than some antipathy to coal, oil, and natural gas. Whatever the explanation for why people think solar and wind are cheap, it is not related to their attitudes toward the traditional fuels. In chapter 6, we show that this is mostly just a matter of misinformation. People have formed a belief that these fuels are cheap, probably because the sun and the wind are natural and ever-present. When told the true prices of generating electricity from these sources, they change their attitudes about energy and energy policy accordingly.

Most Americans have opinions about how expensive or clean various energy sources are. Most Americans have a clear sense of the environmental harms and social costs associated with coal, oil, natural gas, nuclear, hydro, solar, and wind power. And, with a very important exception, Americans seem to have the right picture of the costs of using specific fuels to generate electricity. That exception, of course, is that large majorities of people think solar and wind are cheap. While there is some variation in perceptions across different demographic and political groups, the differences across people are unimportant compared to the true drivers of people's perceptions of costs and harms. Those true drivers are the technologies themselves.

People have a general sense of the economic and social costs associated with different ways of generating power. How do these perceptions, and misperceptions, inform the choices we make as consumers and voters? We turn to this question next.

5

Why Do People Hate Coal and Love Solar?

Americans' affinity for solar and wind power and their aversion to coal, oil, and nuclear power is puzzling. Solar and wind, at a very large scale, have yet to be realized or shown to be feasible, while coal, oil, and nuclear are staples of our energy system. Perhaps people simply love what they cannot have, and grow weary of the familiar. Even in terms of social and economic costs, the framework laid out in chapter 2, the choice is either a wash or a slight advantage to traditional fuels. Coal and oil are far superior in terms of economics to solar and wind. The social benefits—namely, avoidance of health costs from air and water pollution resulting from using solar and wind—are substantial, but they currently just offset the economic costs of their large-scale use. The comparison with nuclear is more problematic. Nuclear power has minimal traditional pollution costs and provides more reliable electricity at a far lower price than solar or wind. So why do people want a major expansion of solar and a significant reduction of coal? Why do they want a lot more wind and a lot less oil and nuclear?

Public opinion is often viewed as complex, shaped by a rich texture of considerations, such as where someone lives, who a person is, what they believe politically or value morally, or what disasters and international crises made headlines this month. By these tellings, public opinion about energy is not very meaningful. It is episodic and fractured. Our views on oil are not comparable to our views on nuclear or solar power. Any account of public opinion about energy, then, would have to delve into the complex value systems of the American public and the context of any particular moment of time.[1]

We have arrived at a much simpler explanation. The American public's energy preferences are really about two characteristics of the energy system: economic costs and environmental harms. Is it expensive? Will it make me sick? So far, we have shown that people like alternative fuels and that they think alternative fuels are cheaper and cleaner than the traditional fuels. These results suggest that perceived harms and costs associated with fuels explain why people support some ways of generating power and oppose others, but the link is only suggestive. We have not yet shown that environmental harm and economic cost explain support for those fuels, and are in fact the primary drivers of public attitudes. To do so, we must gauge the weight people place on economic considerations, on environmental concerns, and on other factors in choosing what energy sources they want the United States to use in the future. That is the objective of this chapter.

Briefly, our results are as follows.

First, perceived harms and costs are both highly important in explaining why people want to use an energy source.

Second, perceived harms and costs explain approximately 75 percent of the differences in support for fuels, or 75 percent of the explained or systematic variation. Demographics and partisanship explain only a small percent of the total variation in attitudes.

Third, of these two attributes, harms have greater weight in people's thinking. In fact, the weight of perceived harms is at least twice the weight of perceived costs.

Fourth, the weights are very stable over time, even though the levels of costs have varied.

Fifth, the weights vary little across fuels. That is, the weight of harms and the weight of costs are about the same for coal as they are for solar as they are for nuclear, and so on.

The last finding is extremely important because it suggests that there is a common framework within which to understand the energy choices people make. People think about energy generally through a shared lens of costs and harms, and fit particular fuels into that framework, rather than having fuel-specific attitudes and preferences.

Why, then, the bias toward solar and wind? There are, we will show, three facets to the answer. First, people have the prices wrong. People think that these fuels are cheap, and that factors other than cost impede their introduction. Second, people rightly view these fuels as superior to coal and oil in terms of air and water pollution and toxic wastes. Third, and most significant, people give greater weight to environmental considerations than to economic considerations in assessing energy choices and energy policies. The emphasis placed on environmental considerations is more than twice that of economic costs. Environmental concerns are sufficiently important in people's thinking about energy that, even if everyone knew the right cost information, a large majority of Americans would still prefer toward solar and wind.

Other explanations are certainly possible. These fuels are wrapped in symbolism deeply rooted in our culture.[2] Sunshine and happiness go hand in hand. A lump of coal evokes many a bad Christmas morning. And nuclear power is freighted with the baggage of the atomic bomb and the Cold War. These fuels also have a political face. Miners in West Virginia, Pennsylvania, and Illinois, for instance, have long sided with Democrats over labor and union issues, though miners' dissatisfaction with the Obama administration's environmental policies have increasingly driven them to vote Republican. A Democratic president, Obama also wrapped himself in the promise of "green jobs." What region one lives in or a person's age, income, or education might also predict who supports one sort of fuel more than others.[3] Symbolism, partisanship, and demography undoubtedly shape how we view energy. However, we think (and will show) that our Consumer Model provides a much more powerful explanation for the choices people make and the attitudes they hold. These other explanations have some currency, but the Consumer Model accounts for the lion's share of the systematic variation in people's attitudes.

Our aim is not just to demonstrate that people's attitudes about energy depend on their perceptions of costs and harms, but to measure the strength of the effects of economic and social costs on energy attitudes. People want energy to be both cheaper and cleaner, but which factor weighs more heavily in people's thinking will determine how they make the trade-off between economic costs and social costs.

A Framework for Understanding Public Attitudes on Energy

It is helpful to keep in mind the framework introduced in chapter 2, and we return to it briefly here. When policy experts, such as environmental economists, conduct cost-benefit analysis, they first try to enumerate all of the forms of costs and benefits, all of the harms and risks, that might be associated with a particular fuel. These may be described in terms of attributes or variables—specifically, the economic costs (e.g., prices and transaction costs) and the social costs (e.g., health harms and damage to other economic activities external to energy production and use) associated with specific fuels. They then try to calculate the monetary value of the harms and risks so as to put everything in a common metric and to compare directly the value of one attribute versus another. Within the White House Office of Management and Budget (OMB), economists in the Office of Information and Regulatory Affairs evaluate the costs and benefits associated with every proposed regulation to see whether the collective benefits to society outweigh the costs imposed on industry and consumers.

The economic cost and environmental harm, and the trade-off between them, has been a defining feature of recent policy debates about energy. For example, much of the congressional debate about the American Clean Energy and Security Act of 2009, the U.S. House legislation better known as the Waxman-Markey bill, focused on the costs of putting an emissions cap on carbon dioxide. Advocates of the legislation often pointed to the analysis produced by the U.S. Environmental Protection Agency, which estimated the cost of the legislation to the average household to be somewhere in the range of $79 to $146 per year.[4] The Congressional Budget Office placed the cost at about $175 per year (or $15 per month) for the average American household.[5] Opponents pointed to other estimates, which put the costs at closer to $2,000 a year.[6] The debate during the first term of the Obama administration over new Clean Air Act regulations also focused extensively on whether the benefits of the regulation justified the costs of tightening pollution control standards. And the general cost-benefit approach is often criticized for ignoring (or, at least, poorly estimating) the subjective value that people place on protecting the environment or stewardship of resources for future generations.[7]

The average person, we argue, goes through a similar process, though without the intensive study and sophisticated analysis of the OMB economists. People develop a general sense of the economic costs and environmental harms associated with a fuel. People do not need to be chemical engineers, epidemiologists, or economists to have a sense of the overall harms and benefits associated with different ways of generating power. As we saw in the previous chapter, most people in fact have a reasonably accurate sense of the economic costs and environmental harms associated with most fuels. The next step in making choices about energy use or energy policy is for the individuals to weigh, subjectively, whether they value or want more of the economic benefits or the environmental benefits. Individuals do not need to go through the painstaking accounting process of estimating the monetary value in order to express the value that they place on economic costs and environmental harms. An individual knows instinctively what is more important—the costs he or she is willing to bear to avoid environmental harms or the environmental harms he or she is willing to tolerate in order to keep energy prices from skyrocketing. In fact, the subjective value that people place on economic and social costs is precisely what is missing from the assessments offered by the OMB. Formalized cost-benefit methods used to assess which policies to use do not necessarily reflect what people value, and the subjective valuation might lead to a different answer than a balance sheet of monetized costs and benefits.

The simplified model of consumer choice, then, goes as follows. Energy use and energy policy involves choices, such as which fuel we want energy companies to use, whether the government enacts policies that increase the use of certain fuels, or whether to support or oppose the construction of a new power plant nearby. People think about energy as they think about any consumer good. There are attributes of the good that they want—in the case of energy, they want the cleanest and cheapest source of power possible. Individuals value both economic efficiency (low price) and environmental cleanliness (low harm). In economic terms, they have utility over economic costs and social costs associated with energy. Individuals also have beliefs about how expensive and how environmentally harmful different sources of energy are. People's assessment of power sources reflects both their beliefs about the attributes associated with that fuel and the value that they place on those two attributes. If they value one attribute more than another, then when there is a tradeoff or choice to be made, they will favor fuels that score higher on the more highly valued attribute.

It is worth further distinguishing what one might consider a Psychological Model of consumers and an Economic Model of consumers.

Much psychological research emphasizes the fleeting nature of people's preferences. The weight of an attribute, such as cost or harm, is viewed as the "salience" of that attribute, and it is often viewed has highly malleable and changed by short-term fluctuations in information, such as news stories, movies (e.g., *The China Syndrome*), events, or elite debate. Those in media and public relations often have a vested interest in having others believe that what they do changes the way people think, not just what people know. Economists, on the other hand, tend to treat people's preferences as fairly stable. The weights of attributes are steady features of a person's underlying valuation of the good, a utility function in the language of economics. From this perspective, the way people think is fairly stable, and price fluctuations, events, and news only alter what people know or understand about a good. Our results are strongly consistent with the economic view of people's preferences about energy. Repeatedly, we observe that the attributes of cost and harms are stable over time and are similar across fuel types.

The formulation of the analysis can be sharpened with a simple mathematical representation. Suppose the level of a good desired is represented as a variable Y (e.g., use of a fuel, ranging from not use at all to expand a lot). Also suppose that the level of each of two attributes can be represented as variables, say, X and W, such as the costliness of a fuel (from Very expensive to Very cheap) and the harmfulness of a fuel (from Very harmful to Not harmful at all). Then, the level of the variable desired by an individual can be approximated as a linear equation: $Y = a + bX + cW$, where a, b, and c are numbers that determine how cost and harm translate into support for an energy source or energy policy. The coefficients b and c represent the relative weight that people place on the attributes.

Our goal is to estimate the coefficients or weights that characterize the energy choices people make. This will allow us to determine how attributes of energy translate into support for specific energy sources and energy policies. It will also allow us to answer key questions regarding the nature of people's preferences and public opinion.

Are costs and harms important? If so, we expect to observe large values of the coefficients b and c in statistical analyses and the cost and harm measures ought to explain much of the variation in preference (Y) across people. Costs and harms are informative because they provide a gauge of how much support for a fuel would change in absolute terms and relative to other fuels with advances in technology. Suppose a widely publicized breakthrough in coal technology greatly diminishes emissions of particulates, sulfur dioxide, mercury, and other pollutants. Firms implement the new technology with only a small increase in cost. We may use the weights to examine how much public support for coal would increase in response to that technological advance. Similarly, we can explore public receptivity to new technologies that reduce either the cost or the pollution associated with any technology.

Do people think about all energy sources the same way, or do they treat some fuels differently from others? If they think about them similarly, then costs and harms should have comparable effects on support for all power sources. That is, the coefficients on costs (*b*) ought to be the same for all fuels, and the coefficients on harms (*c*) ought to be the same on all fuels. If people love solar power no matter what, then costs and harms will be uncorrelated with support for solar; that power source would simply be supported no matter what. But different patterns might hold for natural gas or coal.

Do costs and harms offer the main explanation for energy preferences, or are other factors more important? If they are the main explanation, we expect that the differences in preferences for fuels will be explained mostly or entirely by perceived harms and costs. In other words, the baseline levels of support for fuels (the coefficients represented as the intercepts, *a*) ought to be approximately the same after gauging the effects of costs and harms. And we can incorporate other factors, such as partisanship, income, and region, in the analysis to see if those explain a substantial portion of the variation or to see if the relationship between costs and harms is spurious and is in fact an expression of demographic and political factors.

Finally, how does the public value harms relative to costs? How do we make trade-offs of one energy source for another? The coefficients or weights on costs and harms are immediately instructive in terms of how people make this trade-off. Again, it is our view that people do not actually want less economic efficiency or environmental cleanliness. They want more of *both*. But, the weight that they put on each is different, and that differential weight reflects the rate at which they are willing to substitute one attribute for another. As we will show, Americans put more weight on harms than on costs, and they are therefore willing to substitute environmental gains for economic gains.

In the analysis to follow, we estimate a series of multiple regression models to measure the relationship between energy attributes and energy choices. We present the results graphically in the text, but full regression results are included in the appendix (see tables A5.1–A5.4). Of primary interest are the coefficients on two variables, perceived harm and perceived cost, which are the individual-level responses to the items discussed in the last chapter. The models also include variables to reflect the particular fuel in question (the intercepts). The rest of the variables in the model include a large set of individual-level demographic and political characteristics: energy usage, age, gender, minority, education, income, region, and party identification.[8] Past research has shown that attitudes about the environment differ across segments of the population, and it is possible that similar differences are present regarding energy. Among the most consistent predictors of environmental concern are age, education, political ideology, and party identification. Numerous studies have consistently demonstrated that self-identifying Democrats and ideological liberals, in addition to younger and better-educated Americans, tend to express stronger environmental attitudes than do Republicans and ideological conservatives.[9]

Effects of Attributes: The Case of 2002

This project began with a 2002 survey for the MIT study "The Future of Nuclear Power." In that project, we first saw the overwhelming importance of perceived environmental harms and perceived costs in understanding Americans' energy preferences. We were brought on to conduct a survey to examine how much people support nuclear power and explain why people do or do not want to increase use of the fuel. Concerns about global warming motivated the study, and we thought that those concerns, as well as worries about waste and safety, would rise to the top of people's reasons for supporting or opposing nuclear power.

Drawing on the 2002 survey, we presented a series of analyses to our colleagues in the MIT Nuclear Study Group—Paul Joskow, Richard Lester, Ernest Moniz, John Deutch, Mike Driscoll, and Neil Todreas. The key statistical analysis attempted to predict or explain people's preferences regarding the future use of each of seven ways of generating electricity. The dependent variable in these models is an individual's response to the question about how much he or she wants to see the future use of an energy source increased or decreased (these are the energy preferences reported in chapter 3).[10] The predictors or independent variables in this analysis were perceived harms and costs of each fuel, discussed in chapter 4, demographic characteristics of individuals, indicators of where people live (region and urban area), and measures of political party preference.[11] Table A5.1 in the appendix reports the results of separate ordinary least squares (OLS) regression models for each of the seven energy sources.[12]

The public thought much differently about the nuclear problem than we and our colleagues had expected. We were prepared to find deep regional and partisan differences and a strong emphasis on global warming concerns. Instead, we found that perceived environmental harms and economic costs had the largest and statistically strongest effects on people's preferences. The regression coefficients on harm ranged from about 0.35 to 0.5, which implies that moving down one level on the perceptions of harm scale (e.g., from "Slightly harmful" to "Not harmful at all" or "Very harmful" to "Moderately harmful") is associated with about a half- point change on the energy use preference scale. The difference between a fuel that is viewed as very harmful and one that is viewed as not harmful at all is the difference between reducing the use of the fuel a lot and increasing use of the fuel somewhat. This is a very large and substantively important effect. Perceptions of cost were also important, but the effect was considerably smaller than the size of perceived harm.[13]

Perhaps most disconcerting in the results were our findings about global warming. People who said they were concerned about global warming were less, not more, likely to support nuclear power than those who were not concerned. Nuclear power was not alone. Except for solar power, the correlation between concern about global warming and support for a particular fuel was small and insignificant. For the rest, there was no relationship between concern about climate change in support or opposition to the fuel. Rather, the drivers of public attitudes for every fuel were *local* not global: perceived harms from air and water pollution and perceived costs of electricity.

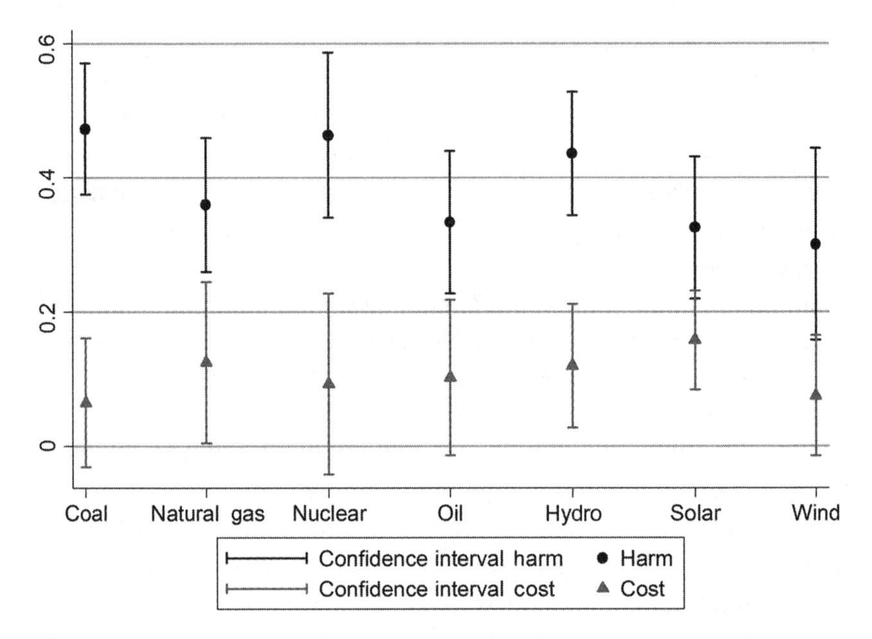

Figure 5.1
Weight of perceived harm and perceived cost on future energy use preferences, 2002 survey
Source: 2002 MIT/Harvard Energy Survey.

More surprising still, the same pattern emerged for every fuel source studied. Perceived environmental harms associated with a fuel had consistently large, negative effects on support for that fuel. That effect dwarfed anything else in the analysis. Perceived cost was a distant second. The coefficient on cost was generally statistically significant, but was about one-quarter to one-half the size of the coefficient on harms. Again, the coefficient was similar across fuels, in the neighborhood of 0.10 to 0.15. Demographics generally did not matter, and when they did they were of secondary importance to perceptions of harm and costs.

Figure 5.1 graphically shows the near uniform effects of perceived harms and costs across the seven types of energy used to generate electricity. This can be thought of as the weight of these attributes on each fuel type. The graph shows standardized regression coefficients and their associated 95 percent confidence interval for each attribute. In this graph, positive values mean that a given attribute is valued more highly.[14] The graph reinforces the conclusion that people place greater weight on harms than costs. The weights attached to harms and costs are generally similar (and statistically indistinguishable) across the fuel types. Moreover, the weights of perceived environmental harms are about the same for most fuels, although they are a little larger when people think about coal and nuclear power.

This highly regular pattern of results led us to think differently about how the public views energy. Americans are not narrowly focused on gas prices or on disasters at oil wells or power plants or such short-term events, as was the emphasis of much of the prior survey research. Public attitudes we found in 2002 were based on a mix of fundamental attributes—environmental harms and economic costs. Americans, it seems, have general opinions about energy, or, in other words, view energy choices through a common lens. The public does not have one set of opinions for nuclear power, and another for solar, and so on. Rather, people want the energy industry and energy policy to deliver a good that has certain attributes. They want electricity, heat, and transportation at a low price and with minimal environmental harm. People have a sense of how much they value each of these attributes. They learn about the costs and harms associated with various fuels and reach a judgment about U.S. energy priorities accordingly.

When we share these results with colleagues, a question invariably arises. Why is the coefficient on economic costs so much smaller than the coefficient on environmental harms? Costs are costs, whether they show up in our energy bills or on our hospital bills. In fact, one of the common stories among economists might lead one to think the opposite. Prices are much more salient and directly tied to energy use than hospital or pharmacy bills associated with asthma and other ailments. The price of a gallon of gasoline or a monthly electric bill is quite salient for most people. Gasoline prices are everywhere we drive; and electric bills are part of our routine lives. Survey data reveal that people know both their electric bills and local area gasoline prices with a high degree of precision, more than most other prices.[15] Because people are so sensitized to and aware of energy costs, one might reasonably argue that such costs should have more weight than health effects of pollution in people's thinking about energy.

We think the explanation reflects the nature of the energy market, and more precisely what people can and cannot get through the private enterprise system. Energy markets are crushingly efficient at delivering reliable power at low cost, especially since the deregulation of the 1980s and 1990s. But energy markets are not as responsive to externalities, such as air and water pollution, since those costs are not reflected in energy prices. Our interpretation of the differential weight of environmental harm and economic cost is that people want more of both attributes—inexpensive power and clean power—but consumer demand for cleaner energy is not in line with the prices that energy firms see. As a result, there is a substantial unmet demand for environmental protection, and that is expressed in public desire for cleaner fuels and in a higher weight on environmental harm in evaluating energy choices.

Stability

Has it always been this way? Was it this way before Silent Spring and Earth Day and the Clean Air Act? We have no way of knowing, but over the years since 2002, we have replicated the study in order to assure ourselves that our initial take was right and to gauge if (and how) the public's attitudes are changing over time. Replication is valuable for two important reasons, one substantive and one methodological. The methodological reason is that one should never put too much faith in a single survey or experiment. Accidents happen; mistakes are made. Indeed, most studies are designed to guard against a 1 in 20 chance of drawing the wrong inference (the statistical power of the study), but your study might be that 1 in 20.

The substantive reason for replication is that public opinion evolves. Over time, people may learn more about energy as the media and elites focus on the challenges facing us.[16] Learning the basic facts about a problem may alter our understanding of the characteristics or attributes of the choices. Information can also alter what we think is important and how we understand the problems we face.[17] It can even alter our basic preferences. Events such as Three Mile Island or Hurricane Sandy may affect people's fundamental valuation of risks and thus the nature of their preferences, not just their understanding of the characteristics of power generation.[18]

The 2002 survey provides a baseline for the rest of the MIT/Harvard Energy Surveys. We repeated parts of the 2002 survey in national representative sample surveys in 2007, 2008, and 2011.[19] The surveys asked the same core set of questions. How costly do you think it would be to provide most of your electricity using each of the following ways of generating power? How harmful to the environment are each of the following ways of generating electricity? Do you think the United States should increase or decrease its use of each of these fuels? And, for each year of the study, we replicated the regression models discussed earlier for the 2002 survey, predicting people's preferences about the future use of each fuel as a function of perceived harms and perceived costs. We also included in these analyses concern about global warming and indicators of political preferences and demographic characteristics.

We performed the analyses separately for each fuel type and year, and the results from 2002 turned out to be typical of the entire decade that followed. Looking year by year reveals that there has been tremendous stability in the factors that have been the main drivers of energy preferences, and stability in the weight of those factors in the way people think about energy choices. Of note, throughout the past decade, perceptions of environmental harms and economic costs largely shape people's assessments of energy choices, and perceived harms weigh about twice as much as concerns about economic costs in that valuation. This is significant given that there have been many high-profile energy-related events—historically high crude oil prices, historically low natural gas prices, the Deepwater Horizon oil spill, the Fukushima Daiichi accident—each of which could have shifted opinion. These events have altered people's assessments of the attributes: the level of cost or the level of harm associated with each fuel. But they have not changed the structure of people's preferences: the weight that people place on harms and costs.

The full statistical analysis is presented in table A5.2 (2007), table A5.3 (2008), and table A5.4 (2011) in the appendix. Figure 5.2 summarizes the key results, showing the estimated weights for local environmental harms, economic costs, and global warming in the assessment of each fuel, over time. Specifically, the figure graphs the standardized regression coefficients from the regressions reported in the appendix. Recall that perceived harms and costs are coded such that positive values reflect views that a fuel is cleaner and cheaper, respectively. Concern about global warming is coded such that higher values indicate greater worry.

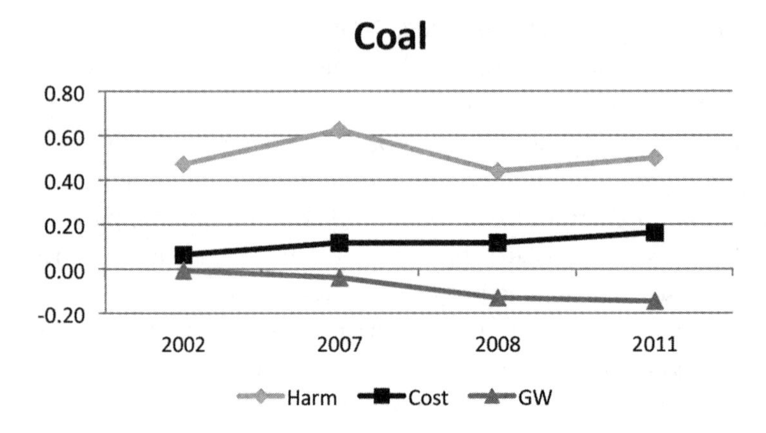

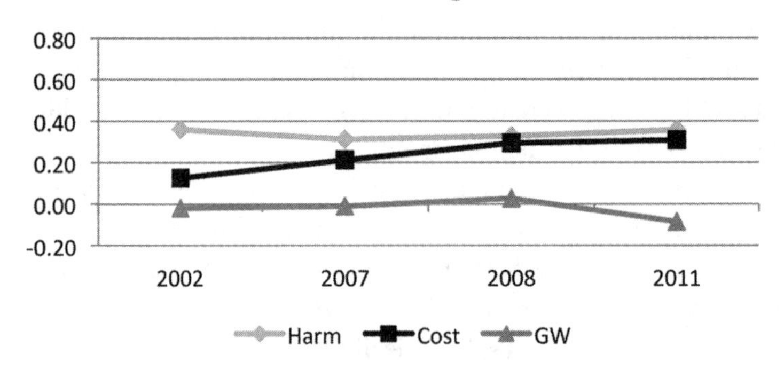

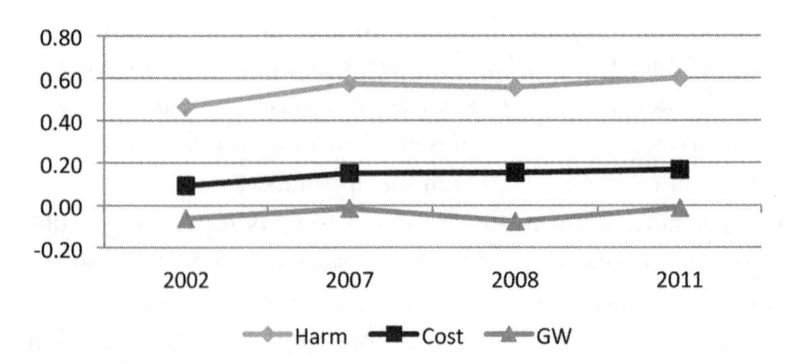

Figure 5.2
Trends in effects of perceptions of harm, cost, and global warming on future use energy sources

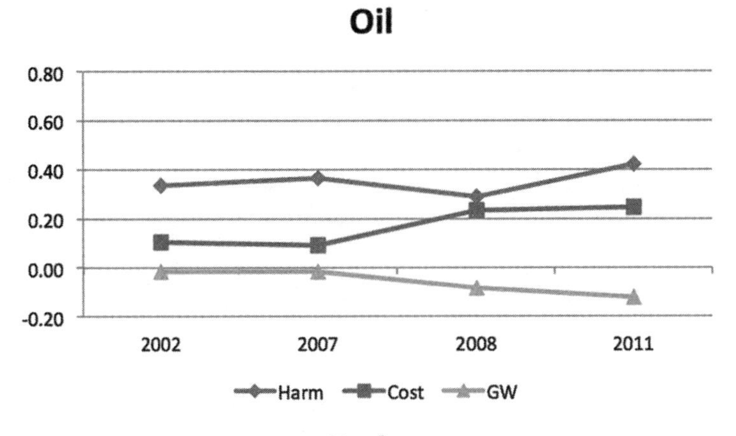

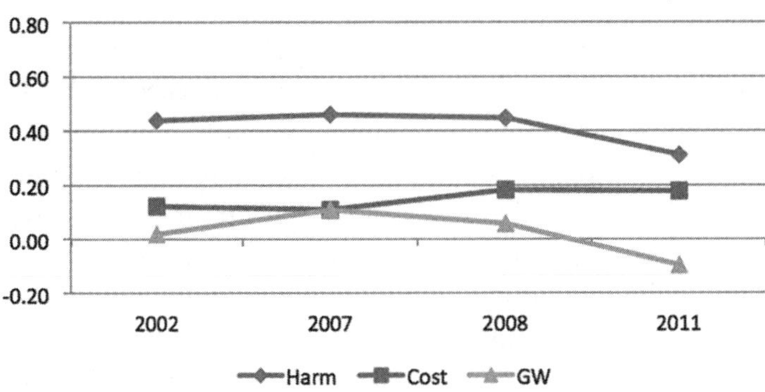

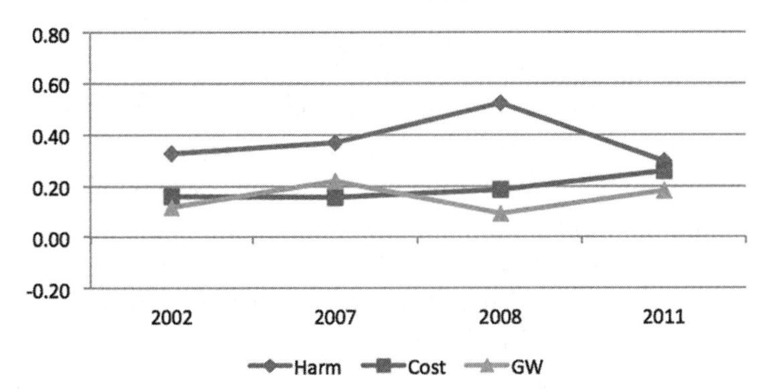

Figure 5.2 (continued)

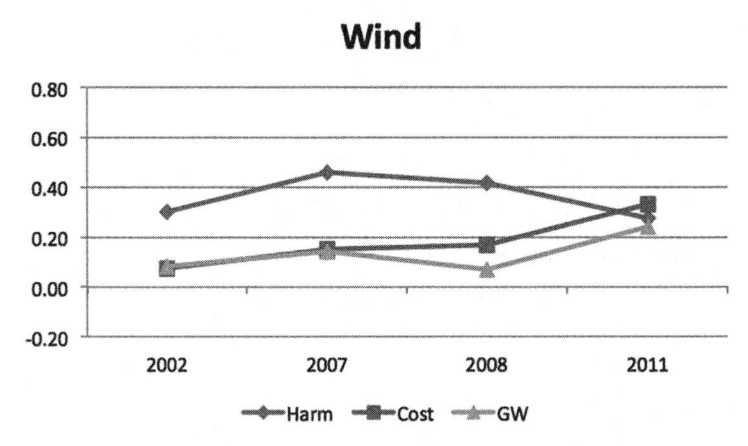

Figure 5.2 (continued)

People's assessments of the environmental impact of energy are the most important factor in how they formulate their preferences. Individuals who evaluate an energy source as environmentally clean prefer to increase its future use to produce electricity. This is true across *all* fuel types and for each year, and the weight of local environmental harms is similar for each fuel type. The coefficients on the perceived harm variable are generally of the same magnitude as in the 2002 survey, as is reflected in the relatively flat lines. There are just two exceptions to this overall pattern. In the 2011 survey, the weight of perceived harms declined for both solar and wind power. Perceptions of harms remain an important consideration in how people evaluate these fuels, but no more so than perceived costs and concern about climate change. We discuss this more below.

Perceptions of cost also remain important. Across the surveys, people who assess an energy source as cheap prefer to increase its future use to produce electricity. This is true whether we are talking about fossil fuels, nuclear power, or renewable technologies. And, as was the case in the 2002 data, in almost every model, the size of the coefficient on perceived costs is much smaller than perceived harm, reaffirming our previous conclusion that people tend to place substantially more weight on environmental impacts than on price.

Oil and natural gas are intriguing exceptions. Throughout the decade there is a significant and steady change in the way people think about these two fuels. The weight of costs becomes increasingly important from 2002 to 2011. In the case of natural gas, by the end of the study period the weight of costs in people's thinking is as important as environmental considerations. Economic considerations, then, play a growing role in the way people think about these two fuels. There is also an increase in the weights of costs in the 2011 survey for solar and wind power. But, for coal, nuclear power, and hydro, the weights of costs held steady for the decade.

The growing weight of economic costs has an important effect on the bottom-line support for oil and natural gas: it accelerates the rising public support for natural gas and opposition to oil. The public perceives (rightly) that natural gas prices have fallen since 2002 and that oil prices have risen. The growing weight of perceived costs means that this attribute has become more important. Hence, natural gas has had a double benefit in the public's mind. It has become cheaper and people have come to think of natural gas more in terms of costs than they did in 2002, and costs now are as salient as environmental harms. For oil, this has been a double whammy. Prices have skyrocketed since 2001, and the salience of those prices has also grown. The growing salience of costs has not come at the expense of environmental harms; perceptions of the harms from natural gas and oil have the same weight they did in 2002.

What about climate change concerns? Here, we see an important change across the spectrum of fuels. The 2002 survey provides a baseline, and in that year concern about global warming registered a significant correlation only with support for solar power, and that correlation was not particularly strong. Since then public discourse about the issue has evolved. The amount of press coverage has grown tremendously, and, unfortunately, the issue has become increasingly partisan.[20] Our analysis further shows that the weight of climate change in people's thinking about energy has risen throughout the decade.

Americans draw the strongest connection between alternative fuels and the need to address global warming. In the 2007 sample, concern about climate change was positively associated with support for solar and wind power; people who were most concerned about global warming were more likely to support the expanded use of these energy sources. Those correlations persisted and strengthened at the end of the study period, as indicated by the increasing size of the effects of global warming on support for solar and wind in figure 5.2.

The disconnect between global warming and fossil fuels persisted throughout much of the decade but began to firmly take hold by 2008. In the 2002 and 2007 surveys, the correlations between concern about global warming and future use of fossil fuels were effectively zero. In 2008, those most concerned about global warming began to exhibit increasing opposition to expansion of coal and oil, and by 2011 global warming concerns at last exhibited statistically significant negative effects on people's preferences about the future use of fossil fuels. Even so, it is important to underscore that the effect is much less than the effects of local environmental harms and economic costs.

Nuclear power remains perhaps the most vexing issue. Nuclear power is widely recognized by climate scientists, economists, and others as a potentially important contributor to any significant and immediate attempt to address carbon emissions. Because it has virtually no carbon emissions, nuclear power has the capacity to offset significant amounts of greenhouse gas emissions from fossil fuels. However, the public doesn't see it that way. As the climate issue has evolved over the past decade, the correlation between concern about global warming and support for nuclear power has remained strongly negative. Today, people who are more concerned about global warming want to use *less* nuclear power. The pattern is not uniquely American. A recent study of British attitudes reaches the exact same conclusion.[21]

This fact exposes a profound political problem for the organization of social movements and interest groups around the global warming issue. Nuclear power is the one non–fossil fuel that can be deployed quickly at an industrial scale to bend the carbon curve in our energy sector. Unlike wind and solar power, nuclear power does not suffer from either the intermittency or the transmissions problems that currently limit these sources, making it a useful way to generate baseload capacity. Nevertheless, most Americans concerned about climate change are not yet on board with nuclear power as a substitute for fossil fuel–based power generation. And the segment of the population that would naturally lead the charge for such a low-carbon source of energy is deeply divided over nuclear power.

This may change in the future as the leadership of environmental organizations grapples with the tricky politics of this issue. Some in the environmental community have moved toward this position, arguing that nuclear power is a necessary part of a future, low-carbon electricity generation portfolio. Patrick Moore, a cofounder of Greenpeace International, used to advocate against nuclear power, as well as nuclear weapons testing. Today, he is working with former EPA administrator and New Jersey Governor Christine Todd Whitman, to advocate for the expansion of nuclear power for the purpose generating electricity.[22] But Moore and other environmentalists who have recently embraced nuclear power reflect a growing split in the environmental movement.[23] Lack of consensus among leaders of environmental groups will make it difficult to organize "concerned citizens" to back nuclear power as a solution for global warming.[24]

Global climate change is, for many, the reason that energy is back on the national agenda. Concern about this issue is driving elite discourse and debate in Washington, D.C.[25] But concern about the climate has not been a primary mover in public attitudes about energy over the past decade. Rather, local environmental concerns and economic costs have had more substantial and immediate impact on public opinion about this subject.

To obtain good estimates of the weight of global warming concerns, as well as the weights of perceived environmental harms and economic costs, we also conducted an analysis that pooled the 2002, 2007, 2008, and 2011 MIT/Harvard Energy Surveys. Combining all years provides more statistical power to measure these effects, and we consider their importance both for preferences about the future use of each fuel as well as the possibility of different types of power plants being located in the respondent's community (discussed in chapter 3). Considering these NIMBY attitudes provides an additional measure with which to assess the relative weights of these factors in people's preferences.

We examined each of the fuel types separately and saw three distinct clusters: fossil fuels (coal, oil, and natural gas), alternative fuels (solar, wind, and hydroelectricity), and nuclear power. The estimates are shown for each of these groups. Table 5.1 presents the estimated effects of harms, costs, and global warming concerns on attitudes about the use of each fuel and NIMBY attitudes toward locating each type of power plant within twenty-five miles of one's home.

Harms and costs are highly important for all three categories of energy (fossil fuels, nuclear power, and alternative energy), and harms have a

Table 5.1
Effects of harm, cost, and global warming concern on use of fuel and NIMBY attitudes, pooled surveys (2002, 2007, 2008, and 2011)

Use fuel*			
	Fossil fuels Coal; oil; natural gas	Nuclear	Alternative fuels Solar; wind; hydro
Perceived harm (1 to 5: High to none)	+0.414 (0.016)	+0.614 (0.025)	+0.303 (0.014)
Perceived cost (1 to 5: Expensive to cheap)	+0.217 (0.016)	+0.103 (0.025)	+0.206 (0.014)
Global warming concern (1 to 4: Not to very)	-0.101 (0.016)	-0.046 (0.024)	+0.230 (0.017)
NIMBY**			
	Fossil fuels Coal; oil; natural gas	Nuclear	Alternative fuels Solar; wind; hydro
Perceived harm (1 to 5: High to none)	+0.125 (0.006)	+0.146 (0.007)	+0.148 (0.011)
Perceived cost (1 to 5: Expensive to cheap)	+0.039 (0.006)	+0.032 (0.008)	+0.025 (0.006)
Global warming concern (1 to 4: Not to very)	-0.064 (0.008)	-0.049 (0.010)	+0.068 (0.009)

*Use fuel has six categories, ranging from "Not use at all" (0) to "Increase a lot" (5). Effects on "Use fuel" are standardized regression coefficients.
**NIMBY takes values Support facility (1) and Oppose facility (0). Regression coefficients are reported.

much bigger effect on people's attitudes than do costs. In all three cases, the difference between thinking a fuel was "Moderately harmful" and "Slightly harmful" (2 units on the scale) translated into the difference between wanting to keep the same level of use for a fuel to wanting to increase use of the fuel slightly (1 unit on the scale). The effect of perceived harms was smallest among alternative fuels (0.3) and largest for nuclear power (0.6). The effect of costs on attitudes and preferences was about half as large as the effect of harms. Prices have a substantial effect on preferences, but that effect is two-thirds the effect of harms in the case of alternative fuels, half the size as the effect of harms for fossil fuels, and one-sixth the size of the effect of harms for nuclear power.

Concerns about global warming have a much smaller effect on people's views about energy than do local environmental harms or costs, but these concerns pull the public toward alternative fuels and push them away from fossil fuels. The net effect of concern about global warming on support for alternatives fuels over fossil fuels is about three-tenths of a point. The effect of concern about global warming on support for nuclear power is negative, and not statistically different from zero.

How Powerful Is the Consumer Model?

Public opinion about energy, we have asserted, largely boils down to just two factors, harms and costs. When we have presented these results to various study groups at MIT, Harvard, Georgetown, and elsewhere, we are often asked two questions. First, what about other factors? Public opinion researchers usually find on any given issue that political and demographic factors, such as party identification, education, age, race, and gender, are extremely good predictors of what people want. Many experts in this area expected there to be deep regional divisions or partisan schisms, especially over coal, nuclear, and solar power. Second, do harms and costs explain the differences in levels of support for fuels? It is entirely possible that perceived costs and harms are strong predictors of Americans' attitudes toward energy sources but only account for a small portion of the difference in overall levels of support for fossil fuels, nuclear power, and alternative energies.

To address these questions directly, we incorporated a large number of demographic and political factors in the multiple regression models. Included were indicators of respondent's age, education, income, gender, race, political party, region of residency, and energy use. We also included in this analysis indicators for different types of fuels. After controlling for demographics, political orientations, and perceptions of attributes of fuels, differences in attitudes toward fuels may still persist, especially if other considerations are at work, such as the symbolic importance of coal or an irrational attachment to solar power.

There are two questions to address. First, how powerful are costs and harms compared to other possible explanations? We examine this question by analyzing the percent of the variation in fuel preferences that is explained by each of three broad factors—the type of fuel, the characteristics of individuals, and the attributes of the fuels (costs and harms). Second, how much of the differences in levels of support for fuels does the Consumer Model account for? To address this, we examine the differences in levels of support across fuel types before controlling for costs and harms and after controlling for costs and harms. In other words, how much of the difference in mean levels of support do cost and harms account for, and how much remains after taking into account how people view the costs and harms associated with fuels?

Turning to the first of these questions, costs and harms are by far the most powerful explanation. Using the pooled dataset (that is, combining the 2002, 2007, 2008, and 2011 surveys), we find that over 80 percent (81.6 percent) of the systematic or explained variation in preferences about future use of fuels is attributed to perceptions of harms and costs. And of all variables and factors considered, harms were by far the most important. Of the explained variation in preferences about use of the fuels, 72 percent was attributed solely to perceived harms, 10 percent was due to perceived costs; the fuels themselves accounted for another 8 percent of the systematic variation. Concern about global warming, year, region, party affiliation, gender, income, education, age, and race combined explained only about 10 percent of the systematic variation.

NIMBY attitudes showed a very similar pattern. Fully 80 percent of the systematic variation was due to harms and costs, with harms accounting for 73 percent of the explained variation and costs, 7 percent. Year and gender were the next most important, accounting for about 4 percent each. Political party, region, education, and income all proved to be trivial in explaining opposition to the location of power plants within twenty-five miles of a person's home.

Turning to the second question, we see even further evidence that costs and harms account for most of how people think about energy. Let us begin with the differences in support for fuels presented in chapter 3. The scale of the outcome variable ranged from 0 to 5, where 0 meant "Do not use this fuel at all" and 5 meant "Increase a lot." We found that wind and solar on average scored a 4. That is, the average rating was to

Table 5.2
Harms and costs explain why people support solar and wind and dislike nuclear, coal, and oil

	Prefer use of fuel*			NIMBY attitude**		
	Minus average score for wind			Minus % support locate wind		
	No controls	Individual controls only	With harms and costs	No controls	Individual controls only	With harms and costs
Oil	-1.87	-1.79	-0.67			
Coal	-1.58	-1.43	-0.38	-46%	-44%	-10%
Nuclear	-1.24	-1.03	-0.02	-47%	-46%	-13%
Natural gas	-0.91	-0.75	-0.25	-23%	-20%	-3%
Hydro	-0.67	-0.62	-0.29			
Solar	0.04	0.02	0.06			

*Average score on six-point scale.
**Percentage support for locating a power plant within twenty-five miles.

increase use of these fuels. Coal and oil scored somewhere between 2 and 2.5, meaning that the typical respondent wanted to decrease use of these fuels somewhat. A second measure was NIMBY attitudes, where 1 means a person supports the location of a given type of power plant within twenty-five miles of his or her home and 0 means that he or she opposes it. The average value, then, is the percent of people who support location of a power plant near their home.

Table 5.2 presents the differences in support for each type of fuel compared with wind. The average support for wind is 4, meaning "Increase somewhat." The NIMBY question is expressed in terms of the percentage of respondents who would support location of a given type of power plant compared with the percentage who would support location of a wind power plant within twenty-five miles of their home. Wind, then, serves as a comparison group or baseline, and each entry in table 5.2, then, translates into a difference between support for a given fuel and support for wind on a given metric.

The table orders the fuels from lowest level to highest level of support (compared with wind). Oil has the least support followed by coal, then nuclear, then natural gas. Natural gas is exactly in the middle of the scale with an average value of 3, or "Keep the same." Solar and wind are tied at the top of the list of fuels that people want to increase, with hydro close behind. These are the differences in support across fuels, without trying to explain those differences.

Our approach is to try to explain away as much of these differences as possible using a certain set of variables, either (1) demographic and political factors or (2) perceived harms and costs. If harms and costs completely account for people's preferences, then the differences among the fuels should vanish upon controlling for perceived harms and costs. To address the question at hand, we make two comparisons. First, compare the differences in levels of support for fuels that remain after controlling for individuals' demographic characteristics and political attitudes (the second and fifth columns) with the raw differences in support for the fuels (the first and fourth columns). Second, compare the differences in levels of support for fuels that remain after controlling for perceived harms and costs.

Consider some specific examples from the table. The baseline difference (i.e., with no control variables) between support for coal and support for wind was -1.58. Controlling for demographic characteristics and political attitudes shrank the unexplained or remaining differences across fuels to -1.43, only a 9 percent reduction. However, the difference between support for coal and support for wind drops to -0.38 once we controlled for respondents' perceptions of the harms and costs associated with these fuels. In other words, harms and costs explain 75 percent of the difference in support between coal and wind. Perceived harms and costs explain 73 percent of the difference in support between natural gas and wind, 64 percent of the difference in support between oil and wind, 57 percent of the difference in support between hydro and wind, and 98 percent of the difference in support between nuclear and wind. The last fact is particularly striking because it reveals that the difficulty the nuclear power industry faces with public perception owes entirely to perceived environmental harms and economic costs.

Americans' attitudes about energy use are quite clearly about the characteristics of energy sources, not about who the person is or the political divisions in the nation today. Averaging across fuels, 74 percent of the variation in levels of support across fuels is explained by perceived harms and costs. That is, the variation across fuels is almost 75 percent smaller once we have controlled for harms and costs than the raw deviations in the mean levels of support. By comparison, individuals' demographic characteristics and political orientations explains no more than 10 percent of the variation in attitudes toward different fuels.[26]

Attitudes concerning the location of power plants show a similar pattern. Most Americans support the location of wind power plants within twenty-five miles of their homes, but most oppose coal and nuclear power plants, and people are divided about natural gas. Across all our surveys, 76 percent say they would support or strongly support the construction of a wind farm nearby; 53 percent say they would support a natural gas–fired power plant within twenty-five miles; and only 29 percent say they would support either a coal or a nuclear plant within that radius. The difference between the percent supporting wind and the percent supporting natural gas is, then, 23 percent. The differences in the percent supporting a given sort of power plant and a wind power plant within twenty-five miles are shown in the fourth column of the table.

Perceived harms and costs explain nearly all of the NIMBY attitudes toward different fuels. Controlling for demographic characteristics and political orientations did very little to explain NIMBY attitudes toward different fuels. After including harms and costs in our analysis, the remaining unexplained difference in support for a coal plant versus a wind plant was 10 percentage points (as opposed to 46 points). The unexplained difference between natural gas and wind plants was just 3 percentage points (compared to 23 points). And the remaining gap between nuclear and wind was down to 13 points (versus 47 points). Averaging across all fuels, perceived harms and costs explain 78 percent of the differences in NIMBY attitudes across fuels. This analysis, coupled with the previous one regarding general attitudes about energy use, provide further support for our Consumer Model of energy preferences.

Trade-Offs

An immediate implication of the Consumer Model is understanding how people make trade-offs. The Consumer Model reveals not just how consumers and voters make energy choices, but how they make difficult energy choices.

Public opinion research has long understood that there are trade-offs in the energy domain, especially between economic costs and environmental harms. Rapid expansion of coal and other inexpensive forms of energy in the 1970s and 1980s came with serious environmental effects, such as damage to northeastern forests and agricultural yields from acid rain. Regulations that bring about cleaner ways of producing energy often involve higher prices. Survey researchers in past research asked about those trade-offs directly: which is more important, jobs or the environment, energy development or environmental protection? But such questions are often criticized for masking the fact that people do not always see such trade-offs.[27] Our data certainly supports that criticism, as many Americans think that solar and wind power are both cheap and clean. They see no trade-off.

Nonetheless, the Consumer Model is instructive in terms of how people make trade-offs, when they arise. The essence of the model is the weight that each of the attributes has in people's thinking, that is, in their revealed preferences or utility functions. Suppose that two technology advances occur simultaneously. One makes solar cheaper (say, on par with wind), and the other makes coal cleaner (say, on par with natural gas). Which will have a bigger effect on public attitudes? Similar changes in the monetized value of the economic costs and social costs would be involved in each of these hypothetical cases. But, from the public's perspective, the effect on support for coal would be much larger than the effect of support for solar. Why is that?

What we have shown in this chapter is that harms and costs explain most of the differences in attitudes toward different energy sources, that both harms and costs have substantial effects on people's preferences, and that harms have a much larger effect than costs. Specifically, an improvement in the environmental cleanliness of energy production is valued about two to three times as much as an equivalent improvement in the economic cost of energy production. The coefficients on harms in table 5.1 were on the order of about 0.3 to 0.6, depending on the fuel, but the coefficients on costs were in the range of 0.1 to 0.2.

To put the matter another way, trade-offs should not be viewed as absolutes. Rather, they should be viewed in terms of the rate at which people are willing to substitute improvement along one attribute or dimension for improvements along another attribute or dimension. Technological innovation is constantly improving on both dimensions at the same time, as we discussed in chapter 2. That is, technological innovation constantly pushes to make energy cheaper and cleaner. The relative weights are informative in terms of which dimension the public values more strongly. And the results quite clearly point to the environmental harms. Taking the estimates literally, people appear to be willing to substitute at a rate of 2 to 1 between economic costs and the environment. That is, people want improvements in both the environmental and economic aspects of energy, but they value much more strongly environmental improvements over economic improvements.

Returning to our example, the result is somewhat counterintuitive, but a reduction in the environmental harms associated with coal is more valued by the public than an equivalent improvement in the economic costs associated with solar power. This is not a ceiling effect since most people already like solar. Rather it is because people weigh environmental harms, and improvement in those harms, more heavily than they weigh economic costs. Public attitudes about energy, then, are more responsive to environmental improvements than to economic improvements. In that sense, people are willing to trade off economic efficiency for environmental protection.

So where should the government put its research dollars? If we have estimated correctly the weight the public puts on environmental harm and economic cost, the government should spend more on research and development to improve the environmental aspects of coal, oil, natural gas, and nuclear power than to improve the cost of solar and wind. If our estimates of the rate of substitutes between costs and harms are in the ballpark of what the typical person wants, then the government should put $2 into clean coal for every $1 into research on cheap solar, as that would be in line with the relative weight that the public puts on those two values.

By the same reasoning, government subsidies of wind and solar power increase the amount of low air pollution power generation and bring the entire electricity sector more into line with what Americans want. The government, then, can be responsive to the public by simultaneously investing research on less polluting forms of coal and oil power generation and by subsidizing wind and solar deployment, such as through Renewable Portfolio Standards, production tax credits, feed-in tariffs, or other policies.

Environmental harms and economic costs are, by far, the most important factors that explain why Americans support some fuels over others and why they oppose the location of some sorts of power plants near their homes but not others. We began with the observation that Americans strongly support an aggressive expansion of wind and solar, and most respondents also want to reduce our reliance on oil and coal. The explanation, we can now conclude, lies almost entirely with the way people view the environmental harms and economic costs of those fuels. Between 75 and 80 percent of the variation in support for fuels is explained by these two factors.

We do not deny the relevance of the cultural symbolism of various sources of energy, such as coal or nuclear fission, nor do we mean to suggest that demographics and political attitudes are irrelevant. These factors have some influence on the way that people think about energy choices, but they are of secondary importance. There are also other aspects of public opinion about energy that we have not examined. Whatever their relevance, there is not much variation in preferences across fuels left to explain after taking into account how people perceive the costs and harms of fuels.

Public opinion research on energy often focuses on short-term events, such as oil spills, nuclear accidents, and gas price spikes. Media organizations hit the phones (and now the Internet) asking people their attitudes about the latest crisis. The emphasis on such episodes has made public opinion about energy seem fleeting, fractured, and lacking in any rationale. It is not. People have clear, stable opinions about the energy future that they would like to see in the United States. They know what sort of power plants they would like to see developed, and they know why. There is a simple, unifying structure to public opinion about energy, and that is the desire to have an energy system that simultaneously reduces environmental harms and economic costs.

6

The Chicken and the Egg

Perhaps we have things backward. Our account, so far, assumes an order to the way people think about energy. First, they think about the features of the energy sources, especially prices and local pollution, and then they decide whether they want to use more or less of a given fuel based on these characteristics. If a person believes, say, natural gas to be cheaper or cleaner than other fuels, they will want to use more of it than other fuels. Similarly, if people think that nuclear power is more environmentally harmful than other fuels, they will want to use less of it. But, our view might be the reverse of how people really think. Perhaps people like certain fuels and infer that the fuel must be cheaper or cleaner than the alternatives.

There is plenty of marketing artistry to suggest that consumer choice can work the other way. Consider a typical consumer product, such as toothpaste. Toothpaste can be distinguished along many dimensions—taste, cavity prevention, and so forth. Our argument is that people choose their source of electricity (and their toothpaste) because it does what they want it to at the right price. But, consumers' thinking might work in the reverse direction. Perhaps we like a brand of toothpaste because of celebrity endorsements or catchy advertising jingles, or simply because we grew up using the brand. Whatever the reason, the brand itself may make us feel good. And, at the store, we might buy the particular brand because of the good feelings that it evokes. If someone asks us what the price is or whether it cleans our teeth better than an alternative brand, we may likely say that the product is cheaper than and works better than alternative brands as a way of justifying our purchasing decisions. That is certainly a plausible story.

The same story might account for the patterns we see in people's preferences about energy. It may be the case that people like solar power and wind because the sun and wind are things that they see and feel as part of nature, and they like nature. Further, because people generally like solar and wind power, they may say in a survey that those means of generating electricity are cheap and clean. Likewise, people might dislike oil and coal, for whatever reason, and therefore think they are more expensive or dirtier. And, nuclear power might just scare the hell out of them. How do we know that people's perceptions of the prices of electricity from different sources or of the pollution that is generated lead people to want more or less of a fuel used, rather than the other way around?

There is some indication in the data we have already presented that the perceptions of prices and environmental harm are not driven merely by the fact that people happen to favor a particular fuel source. Specifically, our survey respondents on average get the order of environmental harms right; they see coal and oil as more harmful than natural gas, which is, in turn, seen to be more harmful than wind and solar power. In some regards, the respondents seem to understand price too; they view coal as the cheapest followed by natural gas and oil, and then nuclear power. If perceptions of the characteristics of fuels were driven by preferences about which fuels we should use, then we would expect more confusion about prices and harms.

Where the American public errs, and in a big way, is that most people believe solar and wind power are very cheap. This might reflect justification of prices based on which fuels they like, or it may be wishful thinking. Or, the typical survey respondent may just be wrong. Those errors in perception would lead us to take a different approach to analyzing and interpreting survey responses if people were justifying their understanding of prices on the basis of what they like, rather than on some correct or incorrect factual basis. Do the errors in public perceptions mean that we have incorrectly gauged how perceptions of prices affect support for the fuel? Perhaps perceptions of prices and environmental harms matter much less than we claim. If that is the case, then people will continue to like solar and wind power, even if they know the true price.

It is difficult to untangle which comes first—preferences or perceptions, chickens or eggs. But there is a way to see how much prices and pollution indeed drive people's willingness to use more of an energy source. Consider the counterfactual we have just presented. If our positive attitudes toward an energy source make us think it is cheap, rather than the other way around, then we should continue to like that fuel source just as much if we learn that it is expensive (or dislike it if we learn that it is cheap). That thought experiment suggests we do a real experiment in which we provide people with factual information about energy choices. Throughout our survey research, we have conducted a series of experiments that do just that.

The purpose of these experiments is to confirm whether we have interpreted the survey data correctly. There is an important practical and political lesson as well. The survey experiments offer some indication of how a nationwide effort to educate people about the economic and environmental costs of energy use might shift the American public's attitudes about different fuel sources. Would a campaign to educate people about energy lead the public to support the increased use of alternative energies such as wind and solar? How might such a campaign alter the position of nuclear power in the debate over energy and the environment? And what would be the consequences for traditional fuel sources, such as coal and natural gas?

The results of these experiments, as we shall see, confirm the regression analyses offered in the preceding chapter. They reveal a tension in public opinion. Cost considerations push the American public to support the increased use of coal and natural gas and the reduced use of solar and wind power. Local environmental considerations, such as air pollution and toxic wastes, pull Americans to favor substantial increases in solar and wind power, and reductions in coal, oil, and natural gas. Public opinion about the energy choices that the nation faces reflects that underlying tension between electricity prices and the social costs of environmental harms. We go so far as to argue—and our experiments demonstrate—that this trade-off is felt very much in local terms. That is, local electricity prices matter, and local environmental harms matter. Global warming matters little, if at all, with one exception. Information about global warming seems to increase support for nuclear power. And therein lies one of the great challenges for public policy makers who are mindful of the long-term consequences of the choices we face today.

The Experiments

To enhance our ability to draw correct inferences about the relationships between the perceptions of energy attributes and preferences regarding use of energy sources, we designed and incorporated experiments into many of the MIT/Harvard Energy Surveys. Survey experiments entail the random assignment of respondents to different treatment and control conditions, which facilitates the drawing of causal inferences. These types of experiments are widely used in research on the psychological underpinnings of political attitudes and policy preferences.[1] It should be stressed that the purpose of these experiments is to confirm the assumptions we make in interpreting the survey results and the regression analyses relating preferences about fuels to attributes of those fuels. If those assumptions are defensible, then we expect to see that people who were given correct information about prices or environmental harm exhibit similar shifts in preferences to the differences that we observed earlier.

The experiments included in the MIT/Harvard Energy Surveys followed a common, basic design. The surveys first asked people whether they thought it was expensive or cheap to use a given fuel to generate electricity. The surveys next asked people whether they thought it was harmful or not to the environment to use a given fuel source. However, before asking them whether the United States should increase or decrease the use of each energy source for generating electricity (the dependent variable examined in the regression analyses in chapter 5), respondents in the sample were randomly assigned to different groups. Each group then received either information about the costs and/or environmental harms associated with different energy sources, or no information at all. The latter group served as a control group.[2] The analysis of each experiment then compared the average response across the groups, and if there were sufficiently large differences, we could then attribute them to the information itself. The random assignment of respondents to the groups provided protection against concerns that the groups differed in meaningful ways (i.e., in a fashion correlated with energy choice preferences).[3]

Most of our experiments played off the price of electricity. That was certainly the relevant metric for electricity costs. It was also a good way to gauge the value of environmental harms and to place that on the same scale as electricity costs. Many public policy analyses, in fact, seek to measure the value of environmental harms in terms of the implicit costs of pollution, such as the costs of treating asthma and other lung diseases or days of work lost. Specifically, the experiments provided information to respondents about the price of different energy sources. The point of the experiments was to measure the response of the typical person in the sample to reasonable information about prices and environmental harms. How much did providing such information move the public? Did the observed changes in opinions in response to such information mirror the difference between people in terms of their perceptions of costs or of environmental harms? We also compared across different treatment groups as we sometimes varied the content of the information—say, showing nuclear power to be relatively inexpensive or relatively expensive. The experiments, then, provided a second way to measure the effect of perceived costs or perceived environmental harms on support for each energy source.

We expected the experimental manipulations to affect people in two ways. First, providing people with information ought to have changed their beliefs about what are the true economic and environmental costs of electricity production. The framing of the experiments was expressly in terms of the likely effect on the average electricity bill of the typical family if all electricity were generated by coal, by nuclear power, and so forth. That information ought to have changed the beliefs of everyone somewhat, and it ought to have had especially pronounced effects among those who hold views of energy and environmental costs that differ wildly from the information provided. We expected people to adjust their support of a fuel downward if that fuel was shown to be much more expensive than they believe, and upward if an electricity source was shown to be relatively less expensive than they believe. The greatest inaccuracies in the typical survey respondent's perceptions occurred with the prices of wind and solar power. Most people said that it is inexpensive to provide electricity using solar and wind. Hence, we expected sizable shifts in support for these fuels, relative to coal and natural gas, in response to the experimental manipulations on price. We expected more modest average effects of information on health simply because the typical respondent had the correct relative ordering of the health effects of the various power sources. Providing credible information to people about such health effects, thus, should have had relatively modest effects on the average person.

Second, providing information ought to have affected the thinking of all people by increasing the amount of knowledge they have on the subject. The typical person is not highly knowledgeable about electricity production. Most of us are not immersed in the nitty-gritty details of different nuclear reactor designs or different technologies for storing heat or electricity from wind generators. Nor do most of us pay attention to the occasional reports issued by the International Energy Agency, the U.S. Environmental Protection Agency, or other organizations on the externalities generated by various fuels. In such a world, any credible factual information about economic or environmental costs will improve the certainty with which people hold their beliefs, as well as what beliefs they hold. Psychologists and economists have long known that increasing certainty has the effect of sharpening people's judgments. In a setting of relatively little knowledge, people's judgments are clouded by their aversion to risks. Upon receiving information that natural gas is cheap, someone who believes that natural gas is cheap might express even greater support for expansion of natural gas because the information increases their certainty of belief and reduces their sense of risk associated with natural gas prices. Such effects of increasing certainty are usually of secondary importance to the effects of correcting incorrect beliefs. Even still, we expected that providing information about the price or harm of a given fuel might have changed support for a given fuel somewhat even among people who already correctly thought that fuel source to be cheap or clean.

As with any exploration into public opinion, it was possible to examine further how the information affects particular subgroups. The experiments might have been received differently by those who had high electricity bills, those who had low or high incomes, or those who had incorrect beliefs compared with those who had correct beliefs.

Our aim was more modest. We designed the experiments to gauge the accuracy of the more naïve approach offered in chapter 5 to estimate the effects of prices and harms. Specifically, our aim was to show that the changes in people's attitudes in response to manipulation of people's beliefs about electricity prices and environmental harms agreed with the differences in people's attitudes between those who believed a fuel source was expensive and those who believed it was cheap or between those who believed a fuel source was harmful and those who believed it was not.

Electricity Costs

2002 Experiment

The 2002 survey launched this project, yet even then we were mindful of the potential problem of lack of information or of projection of support for a fuel onto beliefs about costs and harms. The idea behind this experiment was to present some people with an accurate picture of energy prices at that time and the likely cost of electricity from each of the possible sources. The control group in this experiment received no information. Those in the treatment group received the following text:

> Alternative energy sources tend to be more expensive. Deutsche Bank, one of the world's leading financial institutions and investors, estimates that coal, natural gas, and oil are the cheapest fuels today. Nuclear power, dams in rivers, and wind power cost about twice as much. And solar power costs five times as much. However, over the next 25 years, the prices of fossil fuels will double. Natural gas, coal, and oil will be as expensive as nuclear power, dams in rivers, and wind power.

Our goal was to see if public attitudes toward each of the energy sources would shift in response to accurate price information. This experiment was designed as part of the MIT Nuclear Study Group, which issued its report "The Future of Nuclear Power" in 2003.[4] A key immediate obstacle to nuclear power according to some in that group was the economic competitiveness of nuclear power in the electricity market. But, at that time, it appeared that rising natural gas, coal, and oil prices would bring electricity production from fossil fuels in line with nuclear power and wind. Our contribution to that analysis was to test what importance existing cost considerations had and whether the future cost scenario would cause people to alter their perspective today.[5]

Table 6.1 presents the results of this initial experiment. The control group consists of those who saw no information, and the treatment group includes those who saw the text describing prices. The outcome variable is the willingness to expand the use of each fuel. Each respondent could choose 0 for "Not use at all," 1 for "Reduce a lot," 2 for

Table 6.1
2002 cost experiment

	Control	Price information		
	Mean	Mean	Effect	Sig.
Coal	2.36	2.45	0.09	
Natural gas	3.04	3.03	-0.01	
Oil	2.22	2.48	0.26	***
Nuclear	2.35	2.50	0.15	
Hydro	3.69	3.46	-0.23	**
Solar	4.38	3.98	-0.40	***
Wind	4.38	4.08	-0.30	***
Observations	452	226		

Significance levels: *$p < 0.10$, **$p < 0.05$, ***$p < 0.01$

"Reduce somewhat," 3 for "Keep the same," 4 for "Increase somewhat," and 5 for "Increase a lot." The table displays the average response. Values less than 3 indicate that people want to reduce the use of an energy source, while values more than 3 indicate that they want to increase the use of that source. The column labeled "Effect" is the difference, reflecting the average effect of the information provided in the various treatments.

This experiment had no statistically measurable effect on support or opposition to coal, natural gas, or nuclear power. It registered a statistically significant increased support for expanding use of oil in electricity production, such that support for oil rose to approximately the same level of support for nuclear power and coal. All three fuels (coal, oil, and nuclear), however, remained relatively unpopular: most survey respondents wanted to reduce the use of coal, oil, and nuclear power slightly, even after learning their true costs. The very small effect of information about prices on attitudes toward traditional fuels is not surprising. These fuels are currently used, and the information provided indicated that they were all roughly the same cost.

Alternative fuel sources were a different story. Compared to the control group, respondents provided information about the relative current and future costs of energy sources were more likely to support the *reduced* use of wind, solar, and hydropower. These differences were each statistically significant. The average treatment effect ranged from about 0.23 for hydroelectricity to 0.40 for solar, meaning that having correct information shifted opinion roughly a quarter to four-tenths of a point on the future use scale. Americans still wanted to expand these fuels significantly, but their support was somewhat more tempered.

These results suggest caution for environmental activists who want to push these fuels aggressively in public debate. The more people learn the true cost of solar and wind, the less support these fuels will have. The effect of such a dialogue would not be enough to make coal or nuclear the preferred option of the American public, but it would be enough to cause people to think twice about the alternative energy sources. Years later, after the 2008 election, the Obama administration implemented heavy subsidies to solar firms hoping to jump-start the green energy industry. Solyndra and Evergreen Solar went bankrupt within two years, and the Obama administration was widely criticized for "picking losers." The 2002 experiment contained an important lesson. Public support for a fuel source does not mean that people ignore the economic costs of the fuel in judging a public policy designed to increase use of that fuel. The public popularity of solar power does not reflect the true costs, and upon learning the true costs, support for solar power weakens. It is not the case, then, that support for solar power leads people to ignore the costs; rather, the costs, when they are realized, are factored into public thinking.

The result of our 2002 experiment, then, predicts a very heavy advantage for incumbent fuels in the electricity sector. New energy technologies must overcome the initial costs of deploying that technology. Early adoptions are expensive and early adopters pay a heavier price. The effects of early adoption of a technology, if not properly managed, will be to increase electricity prices rapidly and in ways that the public becomes immediately aware of. That will have a subsequent effect of lowering public support for the policies required to introduce those technologies.

Important for the sake of our approach in chapter 5, the effects on wind and solar and hydropower are roughly in line with our interpretation of the regressions. Learning information about relative costs decreased support for these fuels by about the amount one would expect given the misperception of most people of the cost of these fuels and our earlier estimates of the marginal effect of cost information on support for these fuels. This initial experiment suggests that our interpretation of the patterns in chapter 5 is sound. But one experiment is not sufficient to allay doubts.

2007, 2008, and 2011 Experiments

We conducted three subsequent experiments to measure the response of consumers to price information. These new experiments were much more direct and explicit than the 2002 experiment. The 2002 experiment presented the cost information verbally, did not offer precise information on the cost of a typical consumer's electricity bill, and included current and future price information. The subsequent experiments provided concrete information about how use of each of the fuel sources had immediate effects on the consumers' electricity bills. We also varied the price information, specifically for nuclear and natural gas. These two fuels deserve close attention as they offer near-term approaches to reducing carbon emissions at a national or global scale. Current nuclear power technology could replace a significant portion of coal-generated electricity worldwide without emitting significant amounts of carbon. Natural gas does emit greenhouse gases but substantially less than traditional pulverized coal technology, and, similar to nuclear, natural gas can serve as an immediate replacement for coal.

In the 2007 survey, we examined the effect of specific price information, and we featured a broad comparison between fossil fuels and alternative fuels as in the 2002 experiment. We also varied the treatments to examine the effect of prices on the choice between natural gas and nuclear power. As a result, there were three conditions in this experiment, a control group and two treatment groups (A and B). About half the sample served as a control group and were provided with no information. A quarter of the sample received treatment A, which showed nuclear power as somewhat less cost effective than natural gas, and a quarter of the sample received treatment B, which presented nuclear power as more cost effective than natural gas, and as cheap as coal. The exact language of the treatments was as follows:

Nuclear price treatment A: The International Energy Agency, the world's leading source of information about energy resources, has estimated the cost of a typical month of electricity for a family of 4 in the US for different power sources.

From cheapest to most expensive their estimates are:

Coal	$100
Natural gas	$125
Nuclear	$150
Oil	$200
Wind	$250
Hydro	$300
Solar	$400

Nuclear price treatment B: The International Energy Agency, the world's leading source of information about energy resources, has estimated the cost of a typical month of electricity for a family of 4 in the US for different power sources.

From cheapest to most expensive their estimates are:

Coal	$100
Nuclear	$100
Natural gas	$125
Oil	$200
Wind	$250
Hydro	$300
Solar	$400

It is important to note that the only difference between the two treatments regards the costs associated with nuclear power. In treatment A, the cost of generating electricity from nuclear power was said to be $150 per month for a family of four, and in treatment B it was said to be $100.

The 2008 experiment followed a similar design, but this time manipulated the cost of natural gas. As with the 2007 study, half the respondents were assigned to a control group that received no cost information. One quarter of the sample was presented with information showing that natural gas was as cheap as coal, with electricity from either costing approximately $100 per month for a typical family, and one quarter of the sample was presented with information showing gas to be $200, twice the cost of coal-generated electricity. The exact language was as follows:

Natural gas price treatment A: The International Energy Agency, the world's leading source of information about energy resources, has estimated the cost of a typical month of electricity for a family of 4 in the US for different power sources.

From cheapest to most expensive their estimates are:

Coal	$100
Natural gas	$100
Nuclear	$150
Oil	$200
Wind	$250
Solar	$400

Natural gas price treatment B: The International Energy Agency, the world's leading source of information about energy resources, has estimated the cost of a typical month of electricity for a family of 4 in the US for different power sources.

From cheapest to most expensive their estimates are:

Coal	$90
Nuclear	$150
Natural gas	$200
Oil	$200
Wind	$250
Solar	$400

We replicated the experiment once more in 2011. Between 2008 and 2011, much had changed in the economy and the energy sector in the United States. The catastrophic collapse of housing prices and the stock market in 2008 brought on a deep recession. In the midst of that recession, a shift occurred in one of the longest-asked environmental questions in public opinion research—should we do more to protect jobs or protect the environment? For the first time in decades, more Americans said we should protect jobs even at the expense of the environment.[6] The rise of new drilling technologies and extraction of natural gas from shale rock formations led to a drop in natural gas prices and rapid conversion of coal-fired generation to natural gas. These new drilling technologies, commonly referred to as hydraulic fracturing or fracking for short, also brought public backlash against the environmental effects of natural gas extraction.

The experiment sought to capture the possible change in public attitudes about energy and the environment generally and toward natural gas in particular. The information in this experiment characterized the traditional costs to consumers in one treatment, and, in a second treatment, the social costs of each fuel, including the associated health and environmental cleanup costs. Here we focus on the economic costs and revisit the social costs when we turn to environmental harms. One third of this sample received no information and served as a control group. One third was presented information on the current cost of electricity production from each fuel source. The treatment read as follows:

The International Energy Agency, the world's leading source of information about energy resources, has estimated the cost of a typical month of electricity for a family of 4 in the US for different power sources.

From cheapest to most expensive their estimates are:

Coal	$90
Natural gas	$90
Nuclear	$150
Oil	$200
Wind	$250
Solar	$400

These experiments are informative as a further inquiry into both the role of cost in shaping public preferences about energy choices and the specific effect of information about the price of two major ways of powering electricity generation: nuclear power and natural gas. Recall from our discussion of the perceptions of energy costs in chapter 4 that the public tends to misperceive the costs of generating electricity from different sources in fundamental ways. Across the various years of the MIT/Harvard Energy Surveys, people often overestimated the costs of conventional ways of generating electricity, especially in the case of coal and nuclear power. By contrast, people significantly underestimate the current costs of producing power from renewable sources such as wind and solar. The information provided in the treatments essentially flips this order and better reflects the actual costs from each source. Doing this allows us to explore attitudes toward different ways of generating electricity when we do away with people's dissonance about energy costs. Thus, to the extent that the price information influences attitudes toward future use preferences, we would expect that the public would become more supportive of fossil fuels and nuclear power and less supportive of renewable sources.

Table 6.2
2007 private costs experiment

	Control	Treatment				
	Mean	Mean		Effect		Significance
Coal	2.14	2.58		0.44		***
Natural gas	2.87	3.08		0.21		***
Oil	1.79	1.96		0.17		***
Hydro	3.36	3.19		-0.17		***
Solar	4.40	3.80		-0.60		***
Wind	4.36	3.88		-0.48		***
		Treatment A	Treatment B	A	B	
Nuclear	2.63	2.76	2.94	0.13	0.31**	
Observations	615	308	333			

Significance levels: *$p < 0.10$, **$p < 0.05$, ***$p < 0.01$

Table 6.2 shows the results of the 2007 experiment. This experiment provided current information about electricity from all sources but also manipulated the price of nuclear power. For clarity, in table 6.2 we combine the two treatments for all energy sources except for nuclear power, since there were no statistically significant differences between the two treatment groups. The effects of providing correct price information about the various energy sources was to substantially increase support for coal, natural gas, and oil and to substantially decrease support for wind and solar. Compared to the control group, support for the future use of coal increased by about 0.44. Similarly, the average treatment effect for natural gas was 0.21 and for oil 0.17. The cost information also substantially lowered support for the three renewable energy sources considered. Support for solar and wind power dropped the most, about six-tenths and five-tenths of a point, respectively, while support for hydropower dropped by almost two-tenths of a point.

The manipulation of the price of nuclear power increased support for that technology, but primarily among those who saw information indicating that nuclear power is the cheapest fuel source. Treatment A placed the cost of generating electricity from nuclear power at $150 (between natural gas and oil), and treatment B placed it at $100 (at the same level as coal). Comparing the control group and treatment A, support for the increased use of nuclear power rises by a modest amount of 0.13, but that increase is not statistically distinguishable from no effect. In treatment B where the cost of nuclear power is stated as $100, support for nuclear power increases by about a third of a point, compared to the control group—a strong and statistically significant effect. To the extent, thus, that the public misestimates the costs of nuclear power, this would suggest that support for its use would be a little higher if people had a clearer idea of the actual expense of generating electricity from nuclear power, relative to the alternatives. Even so, it is important to remember that a value of 3 means that people would prefer to keep the use of nuclear power at about the same level as today—that is, it does not create a groundswell of support for nuclear expansion.

The 2008 experiment corroborates the 2007 results. Table 6.3 presents the effects of this experiment on support for the various fuel sources, as well as the effect of the manipulation of natural gas prices. The price information provided was identical to the 2007 experiment, except the cost of natural gas is modified between the two treatments. In treatment A, the cost of natural gas was set at $100 (level with coal as the most inexpensive source of electricity generation), and in treatment B it was set at $200 (at the same level as oil, and between coal and nuclear, and wind and solar power).

Table 6.3
2008 private costs experiment

	Control	Treatment				
	Mean	Mean		Effect		Significance
Coal	2.14	2.64		0.50		***
Oil	1.91	2.03		0.12		**
Nuclear	2.92	2.98		0.06		
Solar	4.46	3.77		-0.69		***
Wind	4.44	3.82		-0.62		***
		Treatment A	Treatment B	A	B	
Natural gas	2.82	3.01	2.87	0.19**	0.05	
Observations	623	360	447			

Significance levels: * $p < 0.1$, ** $p < 0.05$, *** $p < 0.01$

The 2008 experiment had effects almost identical to those in the 2007 experiment. Regarding fossil fuels, the average treatment effect for coal was about a half point on the scale, and a much smaller, one-tenth of a point, effect on the same scale for oil. Each of these effects was statistically significant. Respondents receiving information that wind and solar power are comparatively expensive ways to produce electricity were on average far less enthusiastic about these sources, with the size of the effects about the same as in the 2007 experiment. The cost information did not matter for preferences about nuclear power, since nuclear power was priced in this experiment at approximately the value at which we would expect no effect given the 2007 results.

Manipulation of natural gas prices, interestingly, closely resembles the pattern observed with the manipulation of nuclear power prices. The group (treatment A) receiving information that the cost of producing electricity natural gas was $100 on average expressed significantly higher levels of support for the increased use of natural gas. However, for the group (treatment B) told that the cost of generating electricity from natural gas was $200, they were no more likely to support the increased or decreased use of this fuel source than the control group.

The 2011 experiment echoes the three preceding cost experiments. Providing people with information about the current cost of electricity generation from alternative sources increases support for fossil fuels and decreases support for alternative fuels, especially wind and solar. As with the other experiments, the respondents in the control group, who received no information, want to increase substantially the use of solar and wind, and they want to decrease use of fossil fuels, especially coal and oil.

When people are provided with information on the actual economic costs of supplying electricity from each of these means, public attitudes toward the various fuels shift noticeably. Support for coal increases from "decrease somewhat" to "keep the same." Support for natural gas also grows, with most people expressing a desire to increase the use of this fuel—a shift from the beginning of the decade. Solar and wind remain popular, with most people favoring a significant increase in these sources of electricity, but the experimental treatments temper people's enthusiasm, decreasing support for these fuels by about half a point on a five-point scale.

In all four experiments, simply providing price information produced a shift in support for the fuels in line with our earlier findings. There is clear evidence, then, that people evaluate fuels on the basis of the price they pay. If the price is shown to be relatively low or if they believe the price is low, people will support expansion of the fuel. If the price is high, people want to decrease the use of that fuel. These experiments show that people's perceptions of prices affect their preferences and attitudes about what fuels are used to generate electricity. In other words, at least as far as prices are concerned, we have the causality right. People first form perceptions of prices; those, in turn, shape choices.

Environmental Harms

We also sought to address the question of causality as it concerns environmental harms in the 2002 and 2011 experiments. This is a somewhat more difficult task because people, on the whole, understand at least the relative environmental harmfulness of different ways of generating electricity. Any experiment that provides correct environmental information about electricity production, then, will have more modest effects on the average respondent. To the extent that people misunderstand basic facts, they think that nuclear power releases significant amounts of greenhouse gases,[7] and they overstate the environmental harms associated with burning oil to generate electricity.[8]

The 2002 survey sought to estimate the causal effect of information about harms. The sample was randomly divided into five groups. As described earlier, one group (A) served as a control group, and a second group (B) received information about current prices of power generation and projected increases in the costs of fossil fuels relative to other energy sources over the next twenty-five years. The study also included three experimental treatments that contained information about environmental harms in addition to information about costs. One group (C) was provided information about global warming threats from burning fossil fuels as well as price information. The purpose of this treatment was to emphasize the global warming impacts of burning fossil fuels.[9] A fourth group (D) was presented with information about toxic wastes and air pollution from power generation, as well as price information. This treatment aimed to highlight the noncarbon social costs of fossil fuel use. A final group (E) was provided with all three sorts of information—about prices, global warming, and pollution.

The exact language of the treatments was as follows:

B: Price treatment: Alternative energy sources tend to be more expensive. Deutsche Bank, one of the world's leading financial institutions and investors, estimates that coal, natural gas, and oil are the cheapest fuels today. Nuclear power, dams in rivers, and wind power cost about twice as much. And solar power costs five times as much.
However, over the next 25 years, the prices of fossil fuels will double. Natural gas, coal, and oil will be as expensive as nuclear power, dams in rivers, and wind power.
C: Global warming treatment: Alternative energy sources tend to be more expensive. Deutsche Bank, one of the world's leading financial institutions and investors, estimates that coal, natural gas, and oil are the cheapest fuels today. Nuclear power, dams in rivers, and wind power cost about twice as much. And solar power costs five times as much.
There may be other reasons for wanting to use fuels other than coal, oil, and natural gas. Global warming is an increasingly important concern. The consensus in the scientific community is that continued burning of fossil fuels will raise global temperatures by 5 degrees over the next 50 years, causing coastal flooding and droughts.
D: Pollution treatment: Alternative energy sources tend to be more expensive. Deutsche Bank, one of the world's leading financial institutions and investors, estimates that coal, natural gas, and oil are the cheapest fuels today. Nuclear power, dams in rivers, and wind power cost about twice as much. And solar power costs five times as much.
There may be other reasons for wanting to use fuels other than coal, oil, and natural gas. Toxic wastes are an increasingly important concern. A recent DOE study finds that burning fossil fuels, especially coal, produces acid rain and large volumes of mercury, lead, and other toxic substances, which last for millions of years. Alternative fuels, even new nuclear plants, may produce less toxic waste.
E: Combined treatment: Alternative energy sources tend to be more expensive. Deutsche Bank, one of the world's leading financial institutions and investors, estimates that coal, natural gas, and oil are the cheapest fuels today. Nuclear power, dams in rivers, and wind power cost about twice as much. And solar power costs five times as much.
However, over the next 25 years, the prices of fossil fuels will double. Natural gas, coal, and oil will be as expensive as nuclear power, dams in rivers, and wind.
There may be other reasons for wanting to use fuels other than coal, oil, and natural gas.

Global warming is an increasingly important concern. The consensus in the scientific community is that continued burning of fossil fuels will raise global temperatures by 5 degrees over the next 50 years, causing coastal flooding and droughts.
Toxic wastes is another important problem. A recent DOE [U.S. Department of Energy] study finds that burning fossil fuels, especially coal, produces acid rain and large volumes of mercury, lead, and other toxic substances, which last for millions of years. Alternative fuels, even nuclear plants, may produce less toxic waste.

Comparison of each of the treatment groups with the control group (A) offers one way to measure the effects of the information in the treatments. A second comparison of interests isolates the effect of the environmental information. Group B received price information, while groups C, D, and E received price information *plus* environmental information. Comparison of each of these groups with group B offers an estimate of the effect of information about global warming or about toxic waste on peoples' preferences about increasing the use of various energy sources.

Table 6.4 presents the average response for the control group and each of the four treatment groups, where values range from 0 "Not use at all" to 5 "Increase a lot." Again, values less than 3 indicate that people want to reduce the use of an energy source, while values more than 3 indicate that they want the use of that source increased (values of 3 suggest they want to keep it at the same level). The columns labeled "Effect" are the difference, reflecting the average effect of the information provided in the various treatments. The comparison of the price-only treatment with the control group is the same as reported earlier in table 6.1.

As we described earlier, the effect of the cost component of these experiments decreased support for hydro, solar, and wind and increased support for oil. In the aggregate, the environmental harm component of this experiment decreased support for coal, natural gas, and oil, and *increased* support for nuclear power. The global warming information had zero effect on attitudes toward fuel sources, a finding consistent with the regression analyses discussed in chapter 5. Across the seven energy sources, the people provided information about climate change were no different in their energy preferences than the control group. The toxic waste treatment accounts for all of the increase in support for nuclear

Table 6.4
2002 cost and harm experiment

	Control	Price			Global warming (+ Price)			Toxic waste (+ Price)			Combined (Price + GW + TW)		
	Mean	Mean	Effect v. Control	Sig.	Mean	Effect v. Price	Sig.	Mean	Effect v. Price	Sig.	Mean	Effect v. Price	Sig.
Coal	2.36	2.45	0.09		2.35	-0.10		2.36	-0.09		2.13	-0.32	***
Natural gas	3.04	3.03	-0.01		3.03	-0.00		2.99	-0.04		2.77	-0.26	***
Oil	2.22	2.48	0.26	***	2.40	-0.08		2.42	-0.06		2.19	-0.28	***
Nuclear	2.35	2.50	0.15		2.53	0.03		2.71	0.22	***	2.75	0.25	***
Hydro	3.69	3.46	-0.23	**	3.37	-0.09		3.38	-0.08		3.23	-0.23	***
Solar	4.38	3.98	-0.40	***	3.99	0.01		4.01	0.03		4.06	0.07	
Wind	4.38	4.08	-0.30	***	4.00	-0.08		4.02	-0.06		4.02	-0.06	
Observations	452	226			225			224			226		

Significance levels: *$p < 0.1$, **$p < 0.05$, ***$p < 0.01$

power among the environmental treatments. The combined information concerning global warming and toxic wastes lowered support for fossil fuels. This finding suggests that environmental considerations shape preferences about energy use. The results are not terribly sharp, owing, we think, to the vagueness of the treatments concerning toxic waste and other environmental harms (i.e., it is unclear how the treatments map into people's perceptions of environmental harms).

The 2011 survey clarified this issue. By the end of the decade, economists had begun to attempt to monetize the effects of environmental pollution from all energy sources. Building on a paper published by Michael Greenstone and James Looney in the journal *Daedalus*, we developed an experiment that presented people with information about the environmental harms of energy production expressed in terms of social costs—the health care and environmental cleanup costs that we all pay either directly through doctors' bills or indirectly through higher insurance premiums or government regulations or taxes that must be levied to pay for environmental damage.[10] Greenstone and Looney estimated that the true cost of coal—the economic cost plus the social cost—is roughly double what we consumers pay for electricity generated by coal. Monetizing the environmental harms allows us to develop much cleaner experimental comparisons and sharpens the interpretation of any results.

The setup of the 2011 experiment is similar to that of the 2007 and 2008 cost experiments, but in this case we added an additional treatment that presented the social cost as well as the economic cost (or electricity price) associated with each fuel. Specifically, a quarter of the sample received what we will refer to as "traditional cost" information, stating the International Energy Agency's estimates of the cost of a typical month of electricity for a family of four from different power sources, listed in order from cheapest to most expensive. A second quarter of the sample received different cost information—in this case, data on both the "traditional cost" of generating electricity from different sources and the "social costs." The information characterized the traditional costs as those of consumers, and the social costs as the true cost of different energy sources, including the associated health and environmental cleanup costs. The rest of the sample received no information and serves as the control group.

The two treatments were as follows:

Traditional cost treatment: The International Energy Agency, the world's leading source of information about energy resources, has estimated the cost of a typical month of electricity for a family of 4 in the US for different power sources.
From cheapest to most expensive, their estimates are:

Coal	$90
Natural gas	$90
Nuclear	$150
Oil	$200
Wind	$250
Solar	$400

Social cost treatment: The International Energy Agency, the world's leading source of information about energy resources, has estimated the cost to consumers of using different energy sources and the true cost (including health and environmental cleanup) of using different energy sources. For a family of 4 in the US, the true monthly costs of electricity from different power source are:

	Cost you per month	True cost per month
Coal	$90	$200
Natural gas	$90	$100
Nuclear	$150	$175
Oil	$200	$250
Wind	$250	$250
Solar	$400	$400

The fundamental idea motivating this experiment is to provide people with information about the full costs of generating electricity, including those costs that are hidden from them as consumers. Policies such as pollution taxes and cap and trade emissions programs aim to internalize these costs, and ultimately these costs would be passed on to consumers in the form of higher electricity costs. The additional information provided in the "social costs" treatment sends a strong signal that the actual costs of fossil fuels, and a lesser degree nuclear power, are closer to that of renewables (particularly wind) when the full costs of the externalities are taken into account. An alternative way to view this experiment is as measuring the effect of the environmental harm in the same scale as the economic cost—in dollar terms. The difference in the support for specific fuels between the first treatment and the second treatment reflects the willingness to pay higher prices to avoid the health effects of fossil fuels, especially coal.

There are three sets of effects that are of interest: (1) the difference in preferences between the control group and the traditional costs treatment group; (2) the difference in preferences between the control group and the social costs treatment group; and (3) the difference between the traditional costs and social costs treatment groups. We expect a pattern of results for the first two effects similar to before. The third effect provides an additional test, because the difference is in essence a monetized estimate of the health and environmental harms associated with each energy source.

As displayed in table 6.5, the traditional cost information works in a similar fashion as in the earlier experiments. Compared to the control group, those respondents shown just the pecuniary costs, which show coal and natural gas as relatively inexpensive, preferred to increase the use of coal and natural gas. These respondents were also less likely to support the increased use of solar and wind power after learning that these were the most costly ways of generating electricity. The average treatment effect for these renewable was about 0.5 or half a point on the five-point scale. These differences pass conventional tests of

Table 6.5
2011 private and social costs experiment

	Control	Electricity costs			Social costs		
	Mean	Mean	Effect v. Control	Sig.	Mean	Effect v. Elect. cost	Sig.
Coal	2.50	2.85	0.35	***	2.51	-0.34	***
Natural gas	3.34	3.53	0.19	***	3.54	0.01	
Oil	2.31	2.32	0.01		2.34	0.03	
Nuclear	2.81	2.80	-0.01		2.93	0.12	*
Solar	4.08	3.57	-0.51	***	3.72	0.15	*
Wind	4.01	3.54	-0.47	***	3.73	0.20	**
Observations	1022	495			483		

Significance levels: *$p < 0.1$, **$p < 0.05$, ***$p < 0.01$

statistical significance, and they indicate that learning the true cost of electricity from different sources produced roughly a 10 percent increase in support for coal and nuclear power and a 10 percent decrease in support for wind and solar. These effects are in line with the estimated difference in support for a given fuel between those who thought that fuel was inexpensive and those who thought that fuel was expensive, as shown in chapter 5.

Information on social costs registers a substantial decrease in support for coal and increase in support for alternative fuels, especially wind. Comparing the respondents who were shown the current electricity costs from coal and those who were shown electricity costs plus the social costs provides an estimate of the effect of the social costs on support for coal. Those shown the social costs of coal expressed substantially less support for coal. Both groups wanted to decrease the use of coal, but the group shown the social costs associated with air pollution from coal wanted to decrease coal's contribution to the electricity portfolio much more.

We expected little change in attitudes toward natural gas and oil and nuclear power in this experiment because they were shown to be inexpensive relative to other fuels. There is little change in natural gas and oil, but some increase in support for nuclear power. We also expected little or no increase in support for solar power. Even after accounting for the fact that it has no social cost (at least in terms of pollution), solar is still by far the most expensive source of electricity. Interestingly, there is a slight uptick in support for solar, indicating that people respond to the environmental advantages of solar, even when they are framed in economic terms and even though solar remains the most expensive alternative.

Wind offers the most intriguing case in this experiment. After accounting for social costs, wind is competitive with oil, and much closer in cost to coal and nuclear power. That fact produces a significant increase in support for wind among the group of respondents told the social costs of wind and other fuels. Those informed of the social and economic costs of wind want to see the fuel expanded rapidly and more extensively; those informed of the social and economic costs of coal want to see reliance on that fuel reduced. It is worth stressing that the support for a significant expansion of wind power and significant contraction of coal occurs even among people who are told that coal is still less expensive even after including the social costs. This suggests, as we concluded earlier, that the social costs and environmental harms associated with local area pollution have substantial weight in the way people think about energy choices. The weight of pollution on public attitudes is sufficiently great that people appear to be willing to pay somewhat higher electricity prices in order to gain the health and environmental effects associated with less polluting forms of energy.

We set out in these experiments to understand better how to interpret survey data concerning energy choices. Do people's beliefs about costs, harms, and other attributes of energy sources shape their energy choices? Or is it the other way around—people like what they like and say that solar is cheap and clean because they like it? If the latter were true, then information that changes people's understanding of energy prices or environmental harms associated with energy sources would have little or no effect on the choices people make. The experiments in fact demonstrate repeatedly that beliefs about costs and harms drive consumer choice about energy. People's beliefs may be distorted or inaccurate, such as the optimism that most Americans hold about the price of electricity from solar power, but those distortions are not anchored by a desire to use more of that energy source. Inaccurate beliefs appear to be based on lack of familiarity with the technology and a degree of optimism, but they can be corrected by providing people with correct information. When people are informed of the prices and environmental implications of energy sources, public attitudes move in line with that information.

More striking still, the results of these experiments closely mirror the results from the regression analysis discussed in chapter 5. Cost and health considerations associated with energy production are the important factors that shape how people think about energy choices and energy policy. We found public attitudes shift away from an energy source if that source is shown to be relatively expensive and shift toward an energy source if that source is shown to be among the least expensive alternatives. Across all of our cost experiments, support for coal increases noticeably when people are informed of the cost of electricity from this fuel, and support for solar and wind power drops substantially once people find out the costs of these sources of power. The net effect of learning the price of electricity from each of the energy sources is quite dramatic. Informing the public of the electricity prices associated with all fuels leads to a 30 percent rise in support for coal versus solar and wind on a five-point scale that runs from "Increase a lot" to "Decrease a lot." That is, support for coal rises from "Decrease slightly" to "Keep the same," and support for solar and wind falls from "Increase a lot" to "Increase somewhat." In short, learning the facts about the economics of electricity prices leads people to keep things the way they are, which means heavy reliance on fossil fuels.

The effects of pollution push as strongly in the opposite direction. People want a cleaner environment and they want to avoid the health effects associated with air pollution, water pollution, and toxic wastes that are by-products of burning fossil fuels. Informing people of the health effects associated with traditional means of generating electricity leads to significant reductions in support for coal and significant increases in support for solar and wind power.

The push of electricity prices and the pull of environmental harms create a clear trade-off in aggregate public opinion about energy and, thus, in the way the United States approaches energy policy. The desire to keep energy prices low pushes the public toward traditional ways of generating electricity—coal, natural gas, and nuclear power. The value of a clean and healthy environment pulls people away from fossil fuels and encourages them to support less polluting forms of energy—especially solar power and wind. Our experiments show that trade-off in very clear terms. When we provided people with information that reflected the true price of electricity delivery from various energy sources, opinions shifted noticeably in the direction of fossil fuels and away from alternative sources. When we provided people with information that reflected the harms and social costs associated with various energy sources, opinions shifted noticeably away from fossil fuels and toward alternative sources. Public attitudes about energy choices, then, reflect an underlying balance between electricity prices and social costs. If the social costs of a relatively inexpensive energy source are extremely high, then the public might deem it acceptable to regulate pollution even if that means higher energy prices. In the wake of the Fukushima Daiichi nuclear meltdown, Japan decided to close down its nuclear facilities, even though nuclear power provides 20 percent of the nation's electricity. If the social benefits of a relatively expensive energy source are very great, then the public might deem it acceptable to subsidize electricity from that source. The Green Party in Germany grew in popularity and political power over its opposition to nuclear power and pollution from coal and other fossil fuels, and the German government has long subsidized solar power as a way of increasing the use of relatively clean electricity sources.

Climate change represents a new wrinkle in how society balances prices and social costs of energy use. Burning fossil fuels to generate electricity is a major source of human-made carbon dioxide and other greenhouse gases. That fact would seem to tilt the scale heavily in the direction of social costs. Even though the science behind climate change has been relatively well understood for several decades, public attitudes and policies to avert what may be a looming global catastrophe have changed slowly, if at all. Some of our colleagues have suggested that what we need is a massive education campaign. If people learned the truth, it would change their preferences. Former Vice President Al Gore led such a campaign, beginning in 2006, when he created the Alliance for Climate Protection, which had the mission of convincing Americans of the need to put into place effective solutions to climate change, including those that would fundamentally transform electricity production. These types of efforts have been in the environmental community's playbook for decades. Raising the alarm about pesticides, air and water pollution, and other environmental harms is credited by most environmental policy scholars as being instrumental to the enactment of the suite of pollution control statutes of the 1970s. This strategy has worked in the past.

Climate change is different. Public education campaigns have raised understanding of and concern about the issue, but it has proven much more difficult to take the next steps and translate that concern into concrete and wholesale changes in the ways that we produce and consume energy. Our experiments uncovered just such a disconnect. Climate considerations are not a major driver of public opinion about energy. When told that there is a consensus among scientists that using fossil fuels will result in hotter global temperatures, resulting in both flooding and droughts, people are not moved. We first observed that noneffect in 2002, well before Al Gore's movie *An Inconvenient Truth*, which may have made this subject partisan. And we observed the disconnect between global warming and energy preferences before the so-called Climate-Gate scandal, which may have raised doubts about the veracity of the science itself. In sum, we can infer with even more confidence that concerns about climate change are not driving energy choice preferences, particularly with regard to fossil fuels. It is not that people do not care about the environment and do not think about the environment in making energy choices, either at the level of public policy or in terms of local consumption behavior. Our experiments demonstrate that Americans do connect local environmental harms and their health effects to energy choices. But they have difficulty translating their concerns about global warming into a significant change in energy consumption or production.

We think (though this is hard to demonstrate) that for most Americans, and likely for most people everywhere, the consequences of global warming are simply too remote and too far into the future to justify significant changes in public attitudes and policies affecting energy consumption today. The problem, as it is commonly expressed by our economics colleagues, is that significant efforts to address global climate change will bring significant and immediate increases in energy prices today with no corresponding health or environmental benefits to people living today. Pay more, get nothing. This is not a firm basis for an educational campaign that leads to a groundswell of public support for aggressive climate change policies, and we have shown here that people are simply not making the connections between climate change and energy policy. That situation is a very difficult conundrum, because relatively modest actions now will help future generations avoid the risks of potentially catastrophic changes in the climate.

Is there a way forward? Is there a publicly acceptable approach to addressing climate change? We turn to these questions in the next two chapters.

7

Two Minds about Climate Change

It has been four decades since fears about deteriorating environmental quality led to a sufficient groundswell of political support to change the nation's approach to environmental problems. The issues then were air and water pollution, pesticides and toxics, and threats to endangered species. The effects were obvious—rivers catching fire in Ohio, terrible smog problems in many major cities, widespread use of chemicals such as DDT, and declining populations of American biological icons such as the bald eagle and the California condor. The public was sufficiently energized that an estimated twenty million people participated in some way in the first Earth Day activities in April 1970.

It has also been four decades since the Organization of Petroleum Exporting Countries imposed an embargo on oil exports to the United States as punishment for American support of Israel during the Yom Kippur War. The embargo lasted from October 1973 until March 1974. It caused the price of oil to quadruple and resulted in rationing, the passage of the 55-mile-per-hour speed limit, and the creation of the Strategic Petroleum Reserve.[1] The resulting energy price shocks fueled inflation, which by the end of the 1970s reached rates in the double digits, and caused stagnation in economic growth.[2]

Those were the twin energy challenges of the late twentieth century. They have shaped and continue to shape the way Americans think about their energy use and the energy future, and they have shaped the ambitions of every presidency since. Richard Nixon promised in 1974 to implement policies that would eliminate American dependence on imported oil. Barack Obama announced that the goal of energy independence might soon be achievable. But, concerns about pollution and prices are not what have pushed energy back onto the national agenda. Energy is back for a single reason: global climate change. Climate change is the twenty-first century's energy challenge. The risks associated with global climate change are forcing urban planners and city mayors, especially those in low-lying areas, to develop plans for defense and adaptation; these risks are increasingly the focus of corporate strategies and insurance markets, and they are widely debated in nearly every national government in the world. Energy is back not because there is an energy crisis, but because there is the risk of a looming global environmental disaster.

Global warming is a distinctly American problem. The United States ranks as the nation with the second largest greenhouse gas emissions in the world, behind only China. We are under increasing pressure from other countries to take action to lower emissions. Climate worries, however, will not prompt the United States to change its energy portfolio substantially by the middle of the twenty-first century.

American government moves slowly and by consensus, a consensus that begins with the public. So where is the public on the climate issue today? America is of two minds about climate change. We express concern about the problem in the abstract, but neither is the risk of climate change high on our list of public priorities, nor is it a problem for which we are willing to pay more than a small amount. It is a distant, vague concern, one that may not seriously affect generations alive today, and whose worst effects will likely not be felt on American shores. Climate change is something completely divorced from our practical, day-to-day use of energy. Therein lies the problem.

Energy is the climate problem, but climate is not an energy problem.

Mindset I: A Concerned Public

Climate change has been a growing concern of scientists for more than a half a century, and since the 1980s, there has been a near consensus among experts that increased levels of carbon dioxide and other greenhouse gases in the atmosphere would result in a rise in the earth's surface temperature. For most Americans, however, climate change is a relatively new issue. Many point to the heat wave that hit the central and eastern parts of the country during the summer of 1988 as the event that first put global warming on the national agenda. The National Climatic Data

Center estimated that the 1988 heat wave and associated drought led to five to ten thousand deaths and $40 billion of damage (mostly losses to agriculture).[3] In a June 23 congressional hearing before the Senate Committee on Energy and Natural Resources, James Hansen, Director of the NASA Goddard Institute for Space Studies, testified to the probable relationship between global warming and summer heat waves, and more generally alerted the public to the future risks of climate change.[4] Heat waves, droughts in Texas and other southwestern states, and large hurricanes on the Eastern Seaboard now serve as focal events that environmental advocacy organizations use to bring public attention to this issue.[5]

For the past twenty years, Gallup has been tracking public concern regarding a number of different environmental problems. Generally each spring, Gallup asks a sample of Americans whether they personally worry about a host of problems ranging from air pollution to urban sprawl to global warming. Figure 7.1 shows the responses for global warming from 1989 to 2012 (there is a significant gap in the data from 1992 to 1997 when the question was not asked), grouping together the responses for worrying "a great deal" and "a fair amount" and the

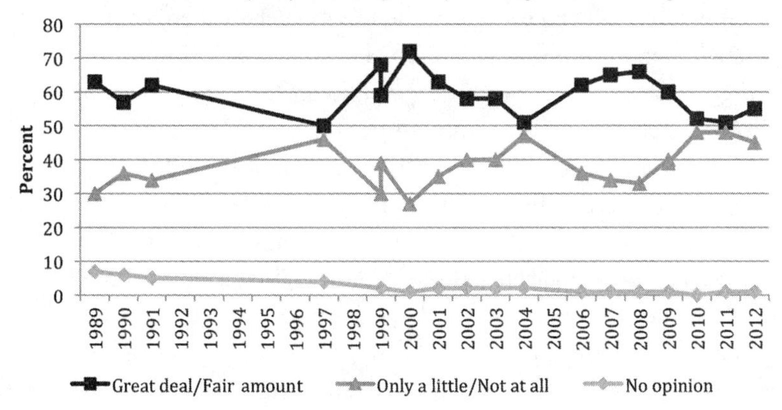

Figure 7.1
Public concern for global warming
Source: Gallup Organization, various years.

responses for worrying "only a little" and "not at all." For the entirety of the time-series, there has been a strong level of concern about global warming.[6]

The MIT/Harvard Energy Surveys show the same high levels of concern about this problem. For example, 85 percent of the public characterized global warming as a very important or somewhat important issue in the 2002 survey. Similarly, the 2003, 2006, 2010, and 2011 surveys asked people about their agreement with various statements about the urgency of global warming. In these surveys, 50–60 percent indicated that they believed that global warming was a real phenomenon meriting some policy response in the near term. The Cooperative Congressional Election Study (CCES) surveys administered annually from 2006 to 2009 with slightly different response categories produced near identical results: approximately 55–65 percent said that global warming is occurring and that some action is necessary. The full question wordings and responses to these items are presented in table A7.1 in the appendix.

Concern about this issue has not been constant overtime. Public concerns were at their highest in the late 1990s, according to the Gallup data, dropped in the early 2000s, surged again in 2006 and 2007, and ebbed after 2008. Recent research from Matthew Kahn and Matthew Kotchen shows that people's concerns about the environment in general, and global warming in particular, are positively associated with business cycles—the worse off the economy, the less concerned people are about these issues.[7] Other research has reached similar conclusions.[8]

Further variation exists within the public on this issue. Anthony Leiserowitz and colleagues at the Yale Project on Climate Change Communication have found what they call "Six Americas" when it comes to Americans' attitudes about climate change.[9] The Six Americas refer to separate segments of the population that think about climate change in distinct ways, and range from the "Alarmed" and "Concerned" to "Doubtful" and "Dismissive" (with two groups fitting in between—"Cautious" and "Disengaged"). In their March 2012 assessment, they characterized 13 percent of the public as "Alarmed," 27 percent as "Concerned," 29 percent as "Cautious," 6 percent as "Disengaged," 15 percent as "Doubtful," and 10 percent as outright "Dismissive."[10] The size of these groupings has stayed fairly stable since their first study in 2008, with some growth in recent years in the proportion of the public expressing skepticism and some drop in the proportion conveying significant worry.

Who comprises these groups? Leiserowitz and his colleagues identify several distinct patterns, most notably two extremes. The "Alarmed" tend to be liberal to moderate Democrats, and are more likely to be middle-aged, college-educated, upper-income women. The "Dismissive" members of the public tend to be conservative Republicans, and are more likely to be high-income, well-educated, evangelical, white men. The other groups tend to be less obviously different in these attributes, although the "Concerned" are more likely to be "Democrats" and the "Doubtful" are more likely to be Republicans. These differences are consistent with other research on the individual-level characteristics associated with beliefs about climate change, particularly the partisan differences.[11] And recent work has shown that the divide between Democrats and Republicans continues to increase when it comes to opinions about the existence, gravity, and causes of global warming.[12]

There has been much speculation about the cause of the partisan divide, including increased segmentation of news consumption and how the issue is framed by the major political parties, industry, and environmental groups.[13] Others have suggested motivated reasoning as an explanation. Motivated reasoning holds that people make sense of problems by relying on a biased set of cognitive processes. In other words, people process information in a way that conforms to predetermined beliefs about reality (even if incorrect), and they dismiss facts or information that conflict with these beliefs. As it relates to partisanship, this would explain why some Republicans predisposed to doubt the veracity of climate change are likely to reject increasingly robust scientific evidence regarding its impacts.[14]

Analysis of our surveys reveals a similar set of individual determinants of concern about climate change. Table 7.1 presents multiple regression estimates using data from several years of the MIT/Harvard Energy Surveys.[15] Specifically, we predicted individual-level concern about global warming using a wide range of demographic characteristics and political attributes, including where people live, their age, gender, education and income level, race, and political party affiliation. Just as in past research, partisan affiliation correlates strongly with global warming concern.

Table 7.1
Individual-level determinants of concern about global warming

DV: Concern about global warming	2002	2003	2006	2007	2008	2011
Age <30	-0.10	-0.11	-0.08	-0.02	0.09	0.20**
	(0.054)	(0.085)	(0.086)	(0.060)	(0.058)	(0.062)
Age >60	-0.06	-0.01	-0.02	0.01	-0.16**	-0.08
	(0.055)	(0.090)	(0.087)	(0.061)	(0.058)	(0.061)
Minority	0.19**	0.01	0.24**	0.03	0.05	-0.12
	(0.051)	(0.082)	(0.079)	(0.055)	(0.053)	(0.059)
Female	0.05	0.17*	0.09	0.14**	0.11*	0.21**
	(0.042)	(0.068)	(0.068)	(0.048)	(0.046)	(0.050)
Education	-0.04**	0.07**	0.12**	0.05**	0.05**	0.07**
	(0.014)	(0.023)	(0.022)	(0.016)	(0.015)	(0.019)
Income	-0.06*	-0.01	0.03	0.02	-0.03	0.05
	(0.023)	(0.037)	(0.037)	(0.027)	(0.025)	(0.026)
Democrat	0.09	0.19*	0.18*	0.27**	0.30**	0.58**
	(0.051)	(0.081)	(0.079)	(0.055)	(0.104)	(0.079)
Republican	-0.31**	-0.24**	-0.61**	-0.45**	-0.38**	-0.66**
	(0.055)	(0.044)	(0.092)	(0.063)	(0.105)	(0.081)
Northeast	0.13	0.38**	0.14	0.21**	0.10	0.01
	(0.065)	(0.105)	(0.106)	(0.075)	(0.072)	(0.078)
South	0.11*	0.20*	-0.19*	0.02	0.05	0.01
	(0.056)	(0.092)	(0.093)	(0.064)	(0.062)	(0.065)
West	-0.01	0.11	0.04	0.21**	0.11	-0.07
	(0.063)	(0.102)	(0.103)	(0.071)	(0.069)	(0.076)
Constant	3.35**	2.88**	3.13**	2.35**	2.46**	3.20**
	(0.082)	(0.131)	(0.133)	(0.096)	(0.124)	(0.108)
Observations	1,320	1,194	1,225	1,232	1,422	1,753
R-squared	0.110	0.070	0.104	0.130	0.152	0.258

Source: MIT/Harvard Energy Surveys.
Note: Cells contain OLS regression coefficients, with standard errors in parentheses. Significance levels: **$p < 0.01$, * $p < 0.05$.

Compared to individuals not affiliated with one of the major political parties, self-identifying Democrats are more likely to worry about climate change, while self-identifying Republicans are substantially less worried. The growing partisan divide on climate change beliefs also emerges from the analysis. Over the course of the past decade, Democrats have become more worried about global warming, while Republicans express less concern or give less weight to this issue. The only other factors that emerge as consistent predictors of concern about global warming are gender and education. Women tend to be more concerned about global warming than men. In terms of education, in 2002 more educated individuals were less likely to say that climate change was an important issue, but in the other samples education was positively associated with climate change concern.

Unlike energy in general, the issue of climate change increasingly divides the public along partisan lines. Most Democrats and liberals today call for immediate action; most Republicans and conservatives prefer to take a wait-and-see approach. That deep division did not exist ten years ago.

Although it is easy to focus on what divides the public, these facts should not distract from the overall lesson. Most Americans agree with the overwhelming scientific opinion that climate change is a real problem. Most Americans express genuine concern about climate change. And most Americans want government and industry to take at least some action to address the sources of the problem as well as to build resilience to the potential consequences.

Mindset II: Other Priorities

Americans have quite a different attitude about climate change when the issue is taken in a broader context. As deep as our concern with climate change may seem, the issue becomes lost when we consider it alongside the many other concerns facing the country.

Gallup has long asked Americans what they think is the most important problem facing the nation, the MIP for short. The MIP provides a long history of comparable items to gauge what is important to the public, and what is a lower priority. Climate change, the environment, and energy all rank low on the list of Americans' priorities today. For example, in Gallup's January 2012 survey, two-thirds of Americans identified the economy as the nation's most important problem,[16] followed in order by dissatisfaction with government (15 percent), healthcare (6 percent), immigration (3 percent), education (3 percent), and a host of other issues mentioned by 1 or 2 percent of the public. In this particular poll, environment/pollution and energy/lack of energy resources were each mentioned by 1 percent of the public (another 1 percent indicated fuel prices, which Gallup included in their economy tally).[17]

The lack of attention to the environment is not unusual. Historically, only a small percentage of people mention the environment. Since the origin of Gallup's MIP question in 1946, it has exceeded 4 percent only once (in 1989, likely a reaction to the Exxon Valdez oil spill that occurred in March of that year).[18] There have been a couple of periods when energy has emerged as an important issue. Notably, during the oil crises of the 1970s, energy was often mentioned by a sizeable percentage of the public. At the height of the second oil crisis in 1979, 20 percent of Americans identified energy as the most important problem facing the country. There was also a spike in attention to energy (although much less intense than the 1970s) during the second half of the 2000s, likely due to the sharp increase in oil prices. These are aberrations, however, and even at their apex, energy and the environment did not rank highly compared to the economy and many other problems.

We included a version of the MIP question in several of the MIT/Harvard Energy Surveys, and the responses reveal a similar pattern. The 2003 survey provided respondents with a list of twenty-two issues and asked them to indicate the three they considered the most important problem facing the country.[19] The top three issues selected by respondents were terrorism (42 percent), the economy (35 percent), and health care (35 percent). Just 9 percent selected the environment, and in terms of rankings, it was thirteenth out of the twenty-two issues listed. We included a similar item four years later on the 2007 survey, although in this version respondents were asked to select the one problem they believed was the most important facing the United States. The top three issues selected were the war in Iraq (14 percent), health care (11 percent), and immigration (10 percent). By comparison, global warming, energy, and air and water pollution were selected by just 4 percent, 2 percent, and 1 percent of the sample, respectively. Thus, taken in isolation, global warming seems important. That is, when asked, a majority of the U.S. public expresses that it is concerned about it. This concern, however, fades away when judged alongside other domestic and foreign issues.

The real test of the importance of climate change in the public's thinking is to compare it with other environmental issues. Even among environmental issues, climate change does not rank highly. Figure 7.2 graphs responses to Gallup's near annual questions about how much people worry about different environmental problems. The graphic displays the percentage of the American public that indicated they personally worry "A great deal" or "A fair amount" for each problem for the years 2000–2012 for ten problems regularly asked about. The data are revealing when it comes to understanding how global warming compares

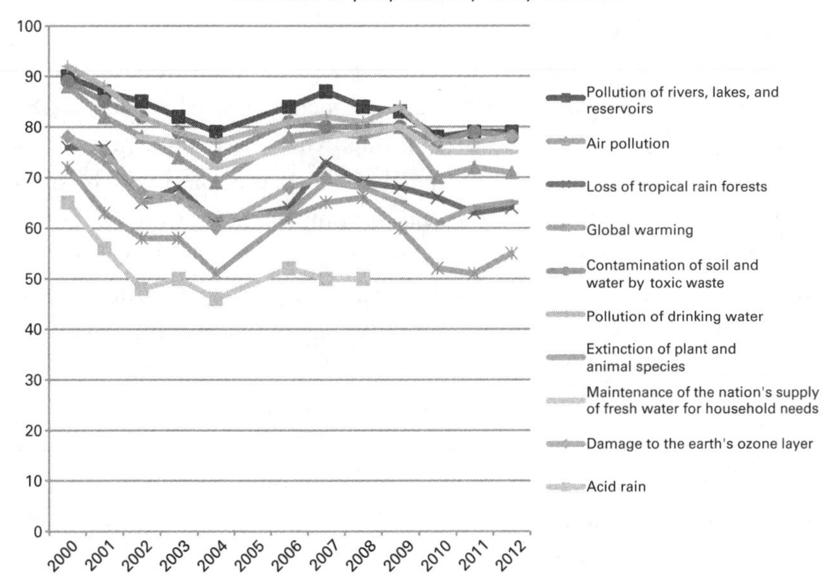

Figure 7.2
Public concern for environmental problems
Source: Gallup Organization, various years.

to other environmental problems. In each survey, people express less concern over global warming than *every* other problem, with the exception of acid rain, a problem that has itself receded in attention over the last decade, to a point where Gallup no longer includes it as part of its annual surveys. Americans are more concerned about local pollution issues, including pollution of rivers, lakes, and reservoirs, air pollution, pollution of drinking water, and toxic waste contamination of soil and water. Even other global issues, such as ozone depletion and loss of tropical rain forests, weigh more heavily on the minds of most Americans than global warming. This is striking given that these issues do not get near the attention they did ten or even twenty years ago.

Perhaps most remarkable in these data is that the gap between global warming and the other problems has grown over the past few years. At the height of public concern about global warming in 2007 and 2008, the gap between the percentage of Americans worrying about this issue compared to pollution of rivers, lakes, and reservoirs was still about twenty points. This gap grew to almost thirty points in 2011. Global warming, according to the Gallup measure, has sunk to the bottom of the list of environmental concerns.

Another way to gauge the priority that people place on global warming is in terms of government attention. We included on the 2007 CCES survey a battery of questions regarding which environmental issues people want the government to spend more time on and which issues people want the government to spend less time on. Specifically, we asked people to evaluate the role of the government in addressing twelve environmental issues: "Thinking about *Environmental Issue X*, how much effort do you think the government should put into addressing this issue?" The response categories were "A lot less," "A little bit less," "About the same," "A little bit more," or "A lot more."[20]

The responses revealed a couple of patterns. First, there was strong support for increasing government attention to each of the twelve issues. Strong majorities (ranging from 53 to 71 percent) of the public supported either "A lot more" or "A little bit more" government effort to address each of the issues; for six of the issues, at least one-third of the respondents indicated that they wanted the government to put forth "A lot more" effort. Second, ranking the responses according to the mean response for the sample reveals that the public prioritizes local and national pollution issues, not global ones like climate change.[21] The top three issues were national or local-level pollution issues. In order they were protecting community drinking water, reducing pollution in U.S. rivers, lakes, and ecosystems, and reducing urban air pollution. Global warming was in the middle of the pack, ranking 8 out of 12, providing additional evidence that for most Americans it is of less concern than many other environmental problems.

These patterns are consistent with our assessment of the importance of local and global environmental issues in public attitudes about energy. People say they are concerned about global warming, but those concerns do not readily translate into priorities. Local environmental harms associated with water and air pollution consistently rise to the top of the public's environmental agenda, not global warming. So how concerned are Americans, and will their concern translate into action?

Are Americans Willing to Pay?

Money is the ultimate measure of how concerned people are. Here we face a true trade-off between economic costs and environmental degradation. There is broad agreement among both advocates and skeptics that addressing climate change in any meaningful way will result in higher energy costs. Moving toward cleaner energy choices will require consumers to pay more for transportation fuels and electricity, at least in the short term. This is true whether the government puts a price on carbon directly through a tax, or indirectly by setting a cap on CO_2 emissions, or by regulating individual sources of these emissions. A key test of the public's resolve and motivation to address climate change, then, is how much Americans are willing to pay.

Economists use the concept of willingness to pay (WTP) to estimate the value or benefits of many environmental policy interventions. WTP reflects the maximum amount that a person is willing to give up to receive a particular good or amenity or to avoid some undesired outcome. Measures of WTP are particularly important when analyzing environmental policies, because unlike most goods and services, environmental amenities (e.g., clean air, clean water) generally are not exchanged in the market. Yet we need to estimate these values in order to evaluate the benefits of various environmental protection measures. Economists have developed two types of techniques to estimate WTP: revealed preferences and stated preferences. The first technique deduces WTP by looking to related markets, and the second asks people directly through a method called contingent valuation. This latter method in essence uses survey questions to ask people what they would be willing to pay to avoid some adverse outcome.[22] These procedures have generated considerable controversy, but they are now routinely applied in environmental economics and in government cost-benefit analyses to evaluate proposed and existing rules.[23]

To further gauge the depth of public concern for climate change, we employed a simple contingent valuation question in several of the MIT/Harvard Energy Surveys. Specifically, we asked respondents to indicate if they would be willing to pay higher energy costs as a way to address climate change. The 2003 survey, for example, included the following question:

If it solved global warming, would you be willing to pay $5 per month more on your electricity bill?

Respondents answering yes were then asked if they were willing to pay $10 per month, with those answering yes then asked about $25, and so on for values of $50 and $100. Respondents who answered no to the first question were assumed to be unwilling to pay anything and assigned a value of zero. It should be noted that the formulation of the question suggests a payment for "solving" global warming, not for an incremental reduction in greenhouse gas emissions or a marginal decrease in temperatures, sea level rise, extreme weather, or some other consequence. The resulting values, therefore, should be interpreted as the additional financial burden people would be willing to bear to address the problem in its entirety. In this sense, they are an extreme upper bound.

Overall, the mean WTP for the public in the 2003 data was approximately $14.50, which equates to about $175 a year. The median was $10. In other words, a majority of Americans said they would be willing to pay at least $10 per month to solve global warming. On average, Americans' electricity bills are about $100 per month, which means that Americans say they are willing to pay about 10 percent more for their electricity to address this problem. As shown in figure 7.3, about 25 percent of the sample was unwilling to pay $5 more a month, and

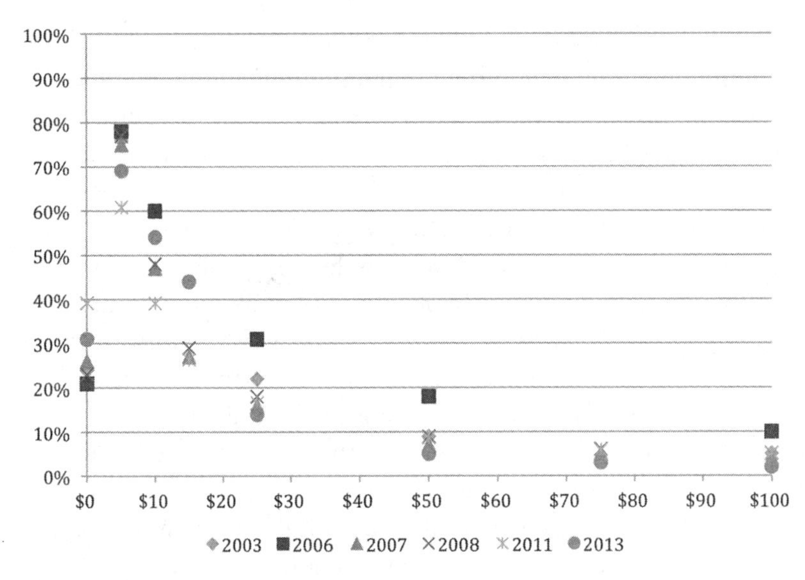

Figure 7.3
Willingness to pay to resolve climate change
Source: MIT/Harvard Energy Surveys.

nearly 80 percent of the sample was willing to pay only $10 or less. Just 9 percent of the public was willing to pay $50 a month, and only 5 percent was willing to pay $100.

Over the subsequent decade, people's willingness to pay higher energy costs for the sake of global warming rises and falls with concern about the problem generally. The 2006 survey included the same series of questions, while the other surveys incorporated a couple of additional payment levels. From 2003 to 2006, the average amount that people said they were willing to pay rose 50 perccent, up from about $14 to $21. This increase, however, was short-lived, as the average dropped back to near 2003 levels in the 2007 and 2008 surveys, and all the way down to about $12 in the 2011 and 2013 surveys. And the median amount fell to $5 in three of the last four surveys, further indicating a decline in willingness to pay.[24] As shown in figure 7.3, over 70 percent of the public in the 2007, 2008, and 2011 surveys were unwilling to pay more than an additional $10 on their monthly electricity bill (56 percent in 2013). And, in the 2011 survey, almost 40 percent of the public was unwilling to pay *anything* (this was over 30 percent in the 2013 survey). Some of this drop certainly reflects the general economic conditions at the time of the surveys, but it does further demonstrate that Americans are, if anything, increasingly disinclined to pay higher energy costs to address the climate problem. Table A7.2 in the appendix shows the data in tabular form.

Our estimates are very much in line with those from other studies. In a recent review of more than two dozen scholarly studies of willingness to pay for climate policy, Evan Johnson and Gregory Nemet found that estimates range from about $27 to $437 per year, with a median value of $135. The underlying studies included in their analysis varied in their geographic jurisdiction (about half were in the United States), policy goal, and payment vehicle, but they collectively suggest a reluctance of the public to pay much to address this problem.[25] In other words, most Americans are not willing to make the trade-off that many policy experts feel is required: substantially higher energy prices in order to substantially reduce carbon emissions.

To put these numbers in context, consider them in relation to the cost estimates of the 2009 energy and climate change bill that the House of Representatives passed. Colloquially known as the Waxman-Markey bill, named after its cosponsors Henry Waxman (D-Calif.) and Edward Markey (D-Mass.), among other features the bill set up a cap and trade system for carbon allowances. Although the bill was contentious for many reasons, one of the most controversial aspects of the legislation was its potential costs. Opponents of the bill frequently referred to it as a "cap and tax." Although ostensibly the costs of compliance would be borne by electric utilities and large industrial sources of greenhouse gas emissions, these costs ultimately would have been passed on to consumers' energy bills. Estimates for these increased costs ranged considerably. The nonpartisan Congressional Budget Office (CBO) placed the average household cost at $175 per year,[26] while analysts at the conservative think tank, the Heritage Foundation, estimated that the costs for an average family of four to be ten times as much ($1,870 per year).[27] Even using the low estimate from the CBO that equates to about $15 a month, our WTP estimates indicate that 70 percent of the public is unwilling to pay an amount two-thirds as much as that which would have resulted under the Waxman-Markey bill.

What explains people's (un)willingness to pay to address climate change? One would, of course, expect that income matters, and multiple regression analysis clearly demonstrates its importance. We estimate statistical models to explain levels of willingness to pay in four of our surveys. Included in the models are the set of explanatory variables that we used to explore concern for climate change (demographics, partisanship, geographical region, and electricity usage). The results from these models are presented in table 7.2. Across many of the surveys, income is positively correlated with willingness to pay, which is consistent with the notion that environmental protection is like any other normal economic good (a good for which demand increases as income increases). A household with income between $100,000 and $124,000 is willing to pay about $2 more per month to solve global warming than a household with income between $40,000 to $49,999 (approximately the median household income in the United States). Education is another important factor, with those with more formal education expressing higher WTP. And, as was the case with concern about climate change, there is an important partisan divide. Republicans had much lower WTP compared to political independents, and the difference between Republicans and Democrats was consistently about $6 to $8 dollars over the course of the surveys. In short, the people willing to pay amounts on the scale that is usually called for to address global warming in a serious way are college-educated, high-income, self-identifying Democrats.

Willingness to pay to avert global warming seems to be especially concentrated in a particular subset of the electorate: those who say they are concerned about the issue and have high income. People who do not think that some steps should be taken now are unwilling to pay more than a couple of dollars, regardless of their income. Only those whose income is above the median and say that we need to act now are willing to pay more than $25 per month. That is a relatively small slice of the American public.

Most Americans express concern over climate change and say that some steps should be taken now to address the matter. At the same time, most Americans are also unwilling to spend more than a few dollars more on each month's electricity bill even "if it solved global warming." We are of two minds—concerned and aware of the problem, but unwilling or unable to act.

Global warming does not appear to be the game changer in the energy domain that many experts feel that it needs to be. Concern about global

Table 7.2
Individual-level determinants of willingness to pay to resolve global warming

DV: WTP on monthly electricity bill

	2003	2006	2007	2008	2011	2013
Age <30	0.21	-4.44*	1.79	7.37**	1.45	0.48
	(1.733)	(2.238)	(1.521)	(1.511)	(2.020)	(1.391)
Age >60	-3.49*	-2.63	-1.16	-1.44	-3.61	-1.92
	(1.761)	(2.133)	(1.458)	(1.434)	(1.928)	(1.310)
Minority	-2.84	0.78	-2.92*	-1.11	-3.70	-1.23
	(1.630)	(1.952)	(1.351)	(1.345)	(1.939)	(1.220)
Female	-2.62	-4.48**	0.87	-3.56**	-4.49**	-2.47*
	(1.347)	(1.695)	(1.157)	(1.165)	(1.585)	(1.088)
Education	0.76	3.23**	1.35**	2.01**	2.18**	0.89**
	(0.447)	(0.548)	(0.375)	(0.385)	(0.598)	(0.303)
Income	1.59*	3.32**	2.06**	1.18	4.38**	1.10
	(0.767)	(0.976)	(0.677)	(0.664)	(0.956)	(0.623)
Democrat	4.30**	0.42	-0.65	6.11*	8.42**	7.23*
	(1.615)	(1.962)	(1.339)	(2.803)	(2.653)	(2.891)
Republican	-1.29	-7.46**	-7.63**	-1.24	-5.04	0.19
	(0.860)	(2.251)	(1.485)	(2.835)	(2.736)	(2.942)
Electricity bill	-0.06	0.78	-0.58	0.41	-0.96	-0.49
	(0.432)	(0.555)	(0.308)	(0.320)	(0.503)	(0.250)
Northeast	5.92**	-2.13	1.54	1.33	4.13	3.21
	(2.077)	(2.663)	(1.860)	(1.840)	(2.478)	(1.825)
South	6.22**	-1.35	-0.72	-0.49	6.53**	1.97
	(1.832)	(2.332)	(1.553)	(1.577)	(1.992)	(1.493)
West	6.24**	2.86	4.59**	2.14	2.62	4.40**
	(2.001)	(2.544)	(1.708)	(1.748)	(2.563)	(1.634)
Intercept	9.27**	8.73	10.91**	1.37	7.40	1.57
	(3.510)	(4.629)	(2.686)	(3.624)	(4.358)	(4.323)
Observations	1,091	1,095	1,101	1,295	689	922
R-squared	0.044	0.093	0.086	0.095	0.183	0.084

Note: Cells contain OLS regression coefficients, with standard errors in parentheses. Significance levels: **$p < 0.01$, *$p < 0.05$.

warming has some effect on public attitudes about energy use, but it is a much lower concern and lower priority than addressing local and immediate environmental problems associated with air pollution, water pollution, and toxic wastes.

It is tempting to hop up on the soapbox and express moral outrage at the average American. Vice President Al Gore, climate scientist James Hansen, writer Bill McKibben, and other powerful voices attempt to stir the passions of the American people, through inspiration or fear or moral approbation. "How dare America imperil the globe by burning fossil fuels with little regard for the global risks," we are admonished. Those scoldings seem to have had some effect on the American public. There is more concern about this issue than at the beginning of century, and, at least in recent years, the more concerned people are the more they support the use of alternative energy and oppose the use of fossil fuels. But that concern may not run very deep. Relatively few Americans are willing or able to pay more than a nominal amount to solve the problem. Such amounts are so small that they would have little effect on the U.S. energy portfolio. Small increases in coal or oil prices will not make alternative energy sources competitive or spur innovation in delivering fossil fuels with lower carbon emissions.

There are also risks associated with the current direction of the public discourse on climate change. Impassioned rhetoric lights a fire under those already concerned. Those alarmed about global warming and with resources to spend are willing to pay much more to address the problem. Public discourse on the issue may also turn off potential supporters. Many choose not to believe the message because of the messengers. Others choose to ignore the problem because it is presented in terms so dire and hopeless that simply thinking about the issue will cause someone to sink into an irremediable depression. And the rhetoric increasingly makes climate change highly partisan and politically divisive, in ways that energy is not.

Americans need to hear pragmatic solutions that fit with our approach to energy generally. We think about energy as consumers. We are motivated by the economic costs and local environmental harms, things we can see and feel and name. We need to think and act locally.

8

What to Do?

In 2006, the California State Assembly passed the Global Warming Solutions Act of 2006, affectionately known as AB (Assembly Bill) 32. It was the most aggressive piece of climate legislation ever adopted by an American legislature. The law, authored by Assembly member Fran Pavley and Assembly Speaker Fabian Nunez, authorized the California Air Resources Board (CARB) to develop and impose regulations to reduce California's greenhouse gas emissions to their 1990 levels by 2020. By 2009 CARB had adopted a series of new rules to achieve a 25 percent reduction in greenhouse gases, including an economy-wide cap on emissions and a market through which firms could trade permits to emit such gases, so-called cap and trade.

California's law looked like the foundation for a new national policy to address global warming through a cap and trade system. Passage of AB 32 followed the creation of the Regional Greenhouse Gas Initiative (RGGI) among ten northeastern states and the adoption of renewable fuel standards in many others. It looked, in 2006, as if the public and the states were willing to tackle the climate issue head on. In early 2009, the Obama administration decided to embrace cap and trade as the centerpiece of its legislative initiatives on energy.

But the moment AB 32 became reality, a movement to repeal it began. California is among the leading states in total oil production in the United States, in 2013 trailing only Texas and North Dakota. It is the second leading consumer of electricity, following only Texas.[1] And it is the nation's leading agricultural producer. The scope of the new rules proposed by CARB made clear the extent of new regulations facing California's industry and the cost to its consumers. Opponents of AB 32 seized the opportunity. They decided to follow a time-honored tradition in California and let the public have a crack at the law before it had a chance to go into effect fully. Assembly member Dan Logue proposed an initiative measure to stop the implementation of AB 32. It was given the number Proposition 23 but referred to by its supporters as the "California Jobs Initiative."

Proposition 23 sought to postpone implementation of AB 32 until California's unemployment rate fell below 5.5 percent. The measure gained enough signatures to get on the ballot for the general election in the fall of 2010. The backers of Proposition 23 certainly had reason for optimism. The California electorate had defeated three recent climate initiatives, and in years when Democrats and liberals won sweeping electoral victories. Californians defeated Proposition 87 in 2006, which called for new taxes on gas and oil, and two measures to promote alternative energy measures, Propositions 7 and 10, failed to pass in 2008.[2] Moreover, 2010 was shaping up to be a terrible year for environmental, liberal, and Democratic causes. California was particularly hard hit by the recession; the state's unemployment rate remained stubbornly high at 12.4 percent. Against that background, holding off on onerous new regulations seemed prudent. Oil companies such as Valero and Tesoro spent tens of millions of dollars to support Proposition 23.[3]

The terms of the debate seemed to pit squarely public concerns about climate change against today's economic and energy problems. Viewed from afar, this was not a debate that the climate advocates would likely win. As we have seen throughout this book, Americans' attitudes about energy are shaped much more strongly by the immediate costs (economic and social) of energy production than by concern about future climate change. Polling by Gallup and other organizations in 2010 showed that the deep economic recession had taken its toll on public support for environmental protection. Heading into the fall election, surveys conducted by the *Los Angeles Times* and by the Field Poll showed the electorate split on the measure, with about 20 percent undecided.[4]

The debate on Proposition 23, however, took an unusual turn. The nature of the debate over this piece of climate legislation is best expressed in California's Official Voter Information Guide. The guide, distributed to all registered voters in the state by the Office of the Secretary of State, provides the arguments in favor of and arguments against each proposition. These pro and anti statements are crafted by advocates on each side of the debate and represent the clearest expression of arguments for and against each proposition, and they typically lay out the direct costs and benefits of a given proposition.

The pro–Proposition 23 strategy followed the same strategy as in earlier campaigns. The statement emphasized the costs of the measure to households and questioned the need for the legislation—after all, California alone could not stop global warming. "California already has a $20 billion deficit and leads the nation in lost jobs, home foreclosures and debt," and the measure would cost each household roughly $3,800 per year in higher energy costs. It encouraged Californians to wait for a better economy before imposing new environmental regulations. The website provided to voters was relatively nondescript: Yeson23.com.

The anti–Proposition 23 statement changed the terms of the debate entirely. The opponents to Proposition 23 did not mention global warming, climate change, or greenhouse gases once. Rather, they tied the defense of AB 32 to the health benefits that would be achieved by lowering conventional air and water pollution if the law went into effect, as well as the development of the clean energy industry emerging in the state.

> TEXAS OIL COMPANIES DESIGNED PROP. 23 to KILL CALIFORNIA CLEAN ENERGY and AIR POLLUTION STANDARDS. . . .
> PROP. 23 is a DIRTY ENERGY PROPOSITION that MEANS MORE AIR POLLUTION and INCREASED HEALTH RISKS—*Vote NO.*

And the website? www.StopDirtyEnergyProp.com.

The defeat of Proposition 23 became one of the bright spots for environmental groups and the Democratic Party in the aftermath of what was otherwise a sobering election. The Democrats' loss of the majority in the U.S. House of Representatives meant that the White House could not continue its push for national climate and energy legislation. But AB 32 had survived. The largest state in the country would move forward with new regulations on greenhouse gas emissions and would implement a market system for lowering emissions in an economically efficient manner.

The subtle lesson from the defeat of Proposition 23 lay in the arguments that ultimately carried the day. Passing and successfully defending climate legislation turned not on the risks and costs of global warming to future generations, but on the immediate social costs associated with air and water pollution from the energy sector. The anti–Proposition 23 forces realized that worries about climate change and greenhouse gas emissions were not sufficient to defeat the proposition. Instead, the fate of AB 32 depended on concerns about energy production in the United States that date back at least to the 1950s and the emergence of the environmental movement. People want a cleaner environment so that they can have better health, and they do not trust energy companies to ensure clean production of energy. The anti–Proposition 23 advocates emphasized the co-benefits of reducing oil, coal, and other fossil fuels, especially reductions in asthma and lung cancers, and they vilified the Texas-based oil companies that supported the proposition. Even though AB 32 was a piece of climate legislation—plain and simple—it was debated as if it were any other piece of environmental and energy regulation. And it was the traditional concerns with local environmental harms that kept the nation's most aggressive climate law alive.

Therein lies the answer to what we consider a fundamental puzzle about public attitudes about climate change and energy policy—and a fundamental political mistake by most advocates of climate policy. The puzzle is just this: A large majority of the American public says it is concerned or very concerned about global warming. Yet efforts to adopt such sweeping policies at the state level, such as in California, are highly contentious and uncertain, and attempts to pass such a law at the national level, such as in 2009, have so far bogged down and failed. Why is there an apparent disconnect between public concern about global warming and energy and climate policy?

The reason, as we shall see in the pages that follow, is that although people say they are concerned about global warming, those concerns don't always translate directly into support for climate policies, such as cap and trade or taxes on carbon emissions. The story of AB 32 offers the kernel of an answer. Cap and trade, carbon taxes, and similar policies are designed and typically presented solely as climate policies. But public concern about global warming is not strong enough or deep enough to carry the day in a legislative debate or public campaign, especially when other concerns, such as jobs and energy costs, are raised. But such policies can have other benefits, such as lowering local air pollution levels or reducing water pollution. When Americans realize the other benefits of an energy policy, as was the case with AB 32, public support widens. The political error of many attempts to pass climate policies in the United States (and likely elsewhere) is to design such policies narrowly around the climate issue and not in a way that obtains other benefits, such as reduced air and water pollution, that people value highly today.

Policy Choices

So far, we have examined what sort of energy future the American public wants and why. A large majority of Americans want substantial expansions of wind and solar power and substantial reductions in our reliance on coal and oil as fuels. The public's preferences about the energy future strongly reflect its belief that wind and solar result in much less pollution than coal and oil. Such environmental perceptions weigh very heavily on public thinking. The public also cares about energy prices, and its preference for expansion of wind and solar appears to reflect excessive optimism that these fuels—at least given current technology—could be used to deliver a substantial fraction of our electricity at prices comparable to coal, oil, and natural gas. Correcting those misperceptions alters Americans' preferences about the nation's energy future somewhat, but the polling data consistently show a strong tilt toward greater emphasis on environmental concerns.

This makes some basic economic sense. Firms that generate electricity from coal, natural gas, or nuclear operate in a very competitive market for power, which exerts a strong pressure to lower electricity prices. If one of those fuels suddenly becomes less expensive, it gains the upper hand in delivering electricity. We have seen this occur in recent years with the shift in use from coal to natural gas in some parts of the country. But the social costs associated with fuels are not reflected in those prices. The electricity market fails, then, to allow people to "buy" cleaner water or cleaner air. There is pent-up demand for these attributes for fuels, and, hence, people give even greater weight to those attributes in public opinion polls, because that is the unmet demand in the marketplace. Government policies, such as regulations and fuel taxes, are designed to force the electricity and transportation fuel markets to reflect these social costs. The Clean Air Act (CAA), the Clean Water Act (CWA), and other environmental regulations have sought to force producers and consumers to internalize the social costs of pollution in their economic decisions.

There is a second sort of social cost, which is even more removed from the electricity market in the United States, and that is the social cost associated with carbon dioxide and other greenhouse gases. Over the past several decades, the scientific community has documented increasing concentrations of carbon and other greenhouse gases in the atmosphere, the amount produced by human activities, and the corresponding change in global temperatures and climate patterns. Projecting those trends forward, atmospheric scientists predict that continued burning of fossil fuels for energy production will drastically alter the Earth's climate. Those forecasting models point to widespread droughts, flooding of major coastal cities, disruption of food supplies, and other consequences by the end of this century. As this problem has become clearer to scientists over the past two decades, the drum beat to act has grown steadily louder.

The policy challenge is how to adjust to lower emissions efficiently, so as to have minimal effects on economic growth and prosperity. A cap and trade system such as the one included in California's AB 32 is one of many ways that governments can reduce the environmental consequences of energy production, agriculture, manufacturing, and other economic activities. Of course, other sorts of policy tools are also available to governments. This chapter explores what *policies* Americans support and why. Surprisingly, people want the form of intervention that economists and other policy experts think is least efficient—regulation, without tradable permits or carbon taxes. The reasons, we argue, is deeply political, not partisan, but political nonetheless.

Broadly speaking, three types of policies are commonly suggested: (1) regulate carbon emissions (either directly through a cap or limit on carbon or indirectly through renewable fuel standards), (2) issue tradable carbon emission permits, as with AB 32 (cap and trade), or (3) tax carbon. Economic analyses point to different degrees of efficiency and distortion associated with the different approaches,[5] but, fundamentally, each is designed to lower the amount of carbon and other greenhouse gases that the private market would produce.

First, consider the regulatory approach. The United States has a forty-year history of regulating air and water pollution under the Clean Air

Act and the Clean Water Act, and the framework established by those acts offers one general approach to climate policy. The Environmental Protection Agency (EPA) or a similar entity could set appropriate limits on emissions of carbon, methane, and other greenhouse gases from cars and other vehicles, from industrial facilities, and from power plants and refineries. For example, the EPA could set facility-level performance standards or mandate the installation of certain types of pollution control technology to reduce CO_2 emissions under the CAA for different industrial sectors. The federal government could also set requirements for a minimum amount of renewable fuels in the energy portfolio—a Renewable Portfolio Standard—as thirty states currently have. The federal government could also impose a mandate, such as the one proposed by former senator Jeff Bingaman in the Clean Energy Standard Act of 2012, which would broaden this requirement to other "clean energy" sources. The government could further raise fuel efficiency standards and impose similar standards on power plants and industrial facilities. All of these policies represent regulatory approaches that would limit the nation's carbon emissions, either directly by restricting emissions from manufacturing, industry, and consumers, or indirectly by requiring that a minimum amount of electricity come from noncarbon emitting sources.

The United States has already started down the regulatory path. In 2007, the U.S. Supreme Court ruled in *Massachusetts v. EPA* that the EPA has the authority under the CAA to classify carbon dioxide as a pollutant and regulate it as such. Under the Obama administration, the EPA has begun the lengthy process of developing the rules required to regulate carbon dioxide and other greenhouse gas regulations. The initial step was to determine whether greenhouse gases threaten public health and welfare, and the EPA determined in December 2009 in its so-called endangerment finding that this is indeed the case for six greenhouse gases. This is the key finding under the CAA that provides the EPA with regulatory authority to act. In May 2010, the EPA and the National Highway Traffic Safety Administration finalized a rule to raise fuel economy standards for motor vehicles from model years 2012–2016, which included a first- time standard for CO_2 emissions. These standards were tightened in October 2012 for model years 2017–2025. In May 2012, the EPA proposed a carbon pollution standard for new fossil fuel power plants, which would in essence require new coal plants to use carbon capture and storage technology. And, finally, the EPA may set standards for existing power plants, as called for by Obama in his June 2013 speech on climate change.

We are at the beginning, not the end, of EPA regulatory activity, and the process of writing complicated rules under the CAA often takes a long time. As an example, in 2011 the EPA proposed two new regulations, the Mercury and Air Toxics Standards and the Cross-State Air Pollution Rule. These rules took approximately ten years to write, and, if Congress and the courts do not overturn them, they go into effect in 2015—some fifteen years after the EPA initiated the process of writing the rules.[6] It is reasonable to expect that carbon regulations will have an equally long gestation.

All the while, the EPA can expect continued legal and political pushback from industries and regions that will be most affected. And some members of Congress are certain to resist such policies as well. In June 2012, Senator James Inhofe (R-Okla.) proposed a bill to overturn the new Mercury and Air Toxics Standards Rule; that effort was turned back in the Senate. The U.S. House of Representatives has twice since 2010 approved a bill titled the Regulations from the Executive in Need of Scrutiny Act, better known as the REINS Act. This bill would require any new major rules and regulations issued by agencies (i.e., those with an annual economic impact of $100 million or more) to secure majority approval of the House and Senate before going into effect.[7] The REINS Act failed to win a majority of votes in the U.S. Senate. However, the idea itself reflects the hostility that some in Congress have to the newfound regulatory authority of the EPA and other agencies. It remains an open question as to whether the EPA will retain its authority to regulate greenhouse gases under the CAA, and an act of Congress or a Supreme Court decision could remove or seriously limit the EPA's power. Many of the greenhouse gas regulations developed during the first term of the Obama administration have so far withstood legal challenges in a federal appeals court,[8] but subsequent action is still possible (and likely).

A second approach is to create a market for greenhouse gas emissions in conjunction with a cap. A cap on greenhouse gases produced by cars, power plants, and other industrial sources is a very inefficient way to reduce emissions. Distortions from caps and regulations arise because the limits apply to all firms equally, but it may be very costly for some firms to reduce greenhouse gas emissions and very inexpensive for others to do so.

Compare two hypothetical firms. Firm A manufactures cement, which is an energy-intensive and greenhouse gas–intensive industrial process. Firm B manufactures plywood, a production process that has much lower energy intensity, lower emissions, and the potential to use forest management methods to offset carbon emissions from production. Firm B can achieve an emissions limit very easily, but firm A must install very expensive new technology. Once firm B reaches the target, it has no incentive to reduce emissions further, even though it could do so at relatively low cost. Firm A, however, struggles to comply with the emissions limit and remain competitive. One might get as much reduction in total emissions if firm B were encouraged to reduce its emissions further, and firm A were shown some leniency by regulators. How can that be accomplished?

The solution, developed at a conceptual level in the 1960s, is to create a market for pollution. As part of the 1990 amendments to the CAA, the United States devised just such a market to deal with acid rain caused by sulfur dioxide emissions from coal plants. The idea was to create a set of allowances equal in total to all pollution that the government would permit, and then let firms buy and sell those allowances. Any firm that had difficulty complying with the limits would gladly pay others to lower their emissions; any firm that could lower emissions at little cost would gladly reduce its emissions in exchange for money from the high-polluting firms. All that one needed to develop was a system of property rights or tradable pollution allowances. That system proved amazingly effective at decreasing sulfur dioxide emissions and reducing acid rain.[9]

A system of tradable pollution permits is also known as cap and trade. The number of permits or allowances issued by the government corresponds to the total limit on all carbon emissions—or the cap. The market for such allowances enables those who can comply easily with this emissions cap to sell their excess allowances, while those who have difficulty complying can buy the extra allowances they need—the trade.

Cap and trade systems, if they can be properly structured, hold great promise. With such a market-based approach, it is possible to implement limits on emissions in a much more efficient way than traditional top-down regulations. Total carbon reductions economy-wide would be the same under a regulatory cap and under a cap and trade regime, but the latter approach eliminates the distortions under top-down regulations that arise from differences in the marginal costs of abatement across firms.

There are a number of nascent markets for tradable carbon allowances. The Europe Union launched its Emission Trading System in 2005.[10] The first market for carbon in the United States is the Regional Greenhouse Gas Initiative (RGGI, pronounced "Reggie"), which was created through an agreement of ten northeastern states: Connecticut, Delaware, Maine, Maryland, Massachusetts, New Hampshire, New Jersey, New York, Rhode Island, and Vermont (New Jersey Governor Chris Christie withdrew the state from RGGI in May 2011). These states have agreed to reduce CO_2 emissions from the power sector by 10 percent by 2018 and have set up a market of tradable carbon permits to accomplish that goal. And the state of California passed AB 32, which authorizes the creation of a cap and trade system and a computerized system for tracking emissions.[11] The initial auction of the California emissions market took place in November 2012.

Cap and trade for greenhouse gases rose to national attention in 2009. The Obama administration endorsed the idea of creating a national system of tradable carbon allowances as part of the American Clean Energy and Security Act, commonly known as the Waxman-Markey bill. Cap and trade was the cornerstone of the Waxman-Markey bill, and in a victory for advocates of a comprehensive climate policy, that bill passed the U.S. House of Representatives in the summer of 2009, at about the same time that the House passed its version of Obama's other main agenda item, health care reform. Given the magnitude of the two bills and their implications for certain key states, it looked as if the administration only had hopes of passing one of them. The Obama team chose to push health care, and cap and trade was put on the back burner.

A third approach to climate policy is to tax carbon and other greenhouse gas emissions. To reduce such emissions, the United States could raise the price of using coal, natural gas, and oil through a direct tax on their production or consumption. The idea is that the government would determine the optimal price of carbon, in order to discourage use of fuels that emit carbon into the atmosphere and to encourage use of fuels that do not. This is the simplest and most direct way to adjust the price of energy generated by fossil fuels to account for their social costs.

Carbon taxes themselves take many forms. One approach is to charge an excise or sales tax that is proportionate to the carbon content of the fuel. State or federal regulatory agencies could determine the carbon content of different fuels used for various activities (e.g., home heating oil, gasoline, natural gas–fired electricity) and impose an excise or sales tax on each unit of that fuel that is consumed. In 1990, Finland became the first nation to impose an excise tax on transportation, heating, and electricity based on the carbon content of fuels. Such taxes are very easy to administer, and often go unnoticed by users, as is the case with gasoline taxes in the United States. Currently the U.S. government imposes an excise tax of 18.4 cents on each gallon of gasoline consumed, and states impose further taxes. The tax on an average gallon of gas used in the United States is 50 cents a gallon, about 15 percent of the price at the pump.

Another way to implement a carbon tax is at the point of production. The United States may impose a tax on energy producers and importers that is based on the amount of carbon, methane, and other greenhouse gases in the fuels they have extracted and emissions from the extraction process—such as a "wellhead" tax on oil and gas producers that is proportionate to the greenhouse gas content of the fuels. A tax at the point of production would give companies an incentive to reduce emissions that occur during mining or drilling. This may be a particularly important issue with the extraction of natural gas from shale rock from hydraulic fracturing, as these new drilling technologies can result in substantial methane emissions.

The Canadian province of British Columbia has taken the idea of the carbon tax a step further, designing a tax that is purely regulatory (and not a revenue collection mechanism). In 2008, British Columbia's provincial government adopted *A Budget for Climate Action*.[12] The province imposed a combination of production and consumption taxes by taxing all of the fossil fuel combustion emissions in the region. The amount of the duty was set at $10 per tonne (1000 kilograms) of greenhouse gas in 2008 and increased to $30 per tonne in 2012. The provincial government returned the entire amount of revenue collected by the carbon tax through (1) reductions in small business corporate income tax rates, (2) reductions in corporate income tax rates, (3) reductions in personal income tax rates, and (4) a new low income Climate Act Tax credit.

There are many advantages of carbon taxes over a strictly regulatory approach. Taxes on carbon and other greenhouse gas emissions are more efficient economically than regulation. Specifically, limits involve costly monitoring and enforcement, and they do not create an incentive to reduce emissions further than specified by a limit, even when it is relatively cheap to do so. Regulations would likely obscure the consequences of their behavior from consumers. Taxes, on the other hand, make apparent the cost of our energy choices and the benefits to changing our energy consumption behavior. The salience of taxes can make them politically unpalatable, even if they are a more efficient way than regulations to eliminate externalities.[13]

What Policies Do Americans Support?

As a matter of economics, a straight-up regulatory approach is the least preferred. As a matter of politics, it is the path of least resistance.

Over the past several years, a number of national polling organizations focused their resources on public attitudes about climate policies. From 2007 to 2012, more than twenty different surveys were conducted on this subject by firms and polling organizations such as the Gallup Poll, Princeton Survey Research Associates, Knowledge Networks, and YouGov. They were sponsored by media firms (such as ABC and CBS), by organizations such as the Pew Foundation and the Kaiser Foundation, by corporations such as United Technologies, and by universities, especially teams at Harvard and MIT (led by Stephen Ansolabehere), Stanford University (led by Jon Krosnick), University of Michigan (led by Barry Rabe), and Yale University (led by Anthony Leiserowitz). These efforts have collectively produced a fairly complete picture of the state of public attitudes toward the three general policy approaches we have discussed: (1) limit emissions, (2) tax emissions, and (3) create a market for emissions. Looking across these surveys, a stable and unambiguous picture of public attitudes emerges. Table 8.1 presents the average results of the responses to different types of questions asked by these survey firms about various policies.[14] We report the average percent of

Table 8.1
Public support for climate policies, media and academic surveys 2007–2012

Question	Support (Average)	Oppose (Average)	Number of surveys
Increase alternative fuels (RFS)	81%	18%	2
Stricter pollution regulations	73%	22%	14
Require higher fuel economy	68%	23%	2
Cap greenhouse gas emissions	65%	29%	17
Cap and trade	51%	45%	19
Carbon tax (ranging from $5 to $25)	27%	72%	8

MIT/Harvard Energy Surveys		
2008		
Environmental regulations	Not strong enough: 56%	Too strong: 20%
Carbon tax of $75/month	Support: 15%	Oppose: 56%
Carbon tax with income tax cut	Support: 41%	Oppose: 27%
2010		
EPA cap carbon emissions	Yes: 51%	No: 24%
Carbon tax	Yes: 41%	No: 32%
Cap and trade	Yes: 29%	No: 31%
2011		
EPA cap carbon emissions	Yes: 64%	No: 16%
Carbon tax of $75/month	Support: 25%	Oppose: 49%
Carbon tax with income tax cut	Support: 58%	Oppose: 42%
Carbon tax with deficit reduction	Support: 47%	Oppose: 53%
Clean Energy Act (cap and trade)	Support: 56%	Oppose: 44%

respondents who support and the average percent who oppose each proposal in the table. Most questions allow people to say they are not sure or don't know, so the categories do not total 100 percent.

These surveys produce some striking results. Americans like environmental regulation. They dislike carbon taxes. And they are divided and uncertain about cap and trade.

Regulating greenhouse gas emissions is very popular. In the MIT/Harvard Energy Surveys, regulations routinely receive much more support than energy taxes. For example, the 2011 survey asked people whether they supported or opposed EPA regulation of greenhouse gas emissions. Some 64 percent of people supported an EPA cap on carbon emissions, and only 36 percent opposed that regulation, with the remainder undecided. That compares with 58 percent of people in favor of carbon taxes tied to income tax cuts, and only 25 percent in favor of an outright carbon tax of $75 per month.

Similar differences are reflected in our 2010 survey. A majority (51 percent) of Americans supported EPA regulation of CO_2 emissions from power plants and other large industrial sources, compared to just 24 percent of the public that did not (the rest were either indifferent or answered "Don't know"). Support for a carbon tax was only 41 percent, with 32 percent opposed, while support for cap and trade was less than 30 percent, with about an equal number opposed.[15] Looking at all of the polling data presented in table 8.1, approximately 70 percent of Americans express support for regulations that would cut greenhouse gases, while only 25 percent express opposition to such regulations.

Importantly, support for setting a regulatory cap on carbon and other greenhouse gases mirrors support for *any* sort of environmental regulation. Roughly 75 percent of Americans support stricter regulations of air pollution emissions generally. About 80 percent of Americans support renewable fuel (or portfolio) standards and tighter emission standards "on industry." Two in three support higher fuel economy standards, such as the ones that the Bush and Obama administrations have put in place. Over 70 percent want more money spent on the development of wind and solar power and alternative fuels for automobiles.

Cap and trade is decidedly less popular than outright caps. Two dozen or so surveys conducted by media and academic organizations from 2007 to 2012 asked about support for cap and trade. The topic is sufficiently complex that it was difficult to formulate a simple (or easy) question with which to elicit public opinion. Interestingly, the questions were quite similar across the polls with one question being asked often by many of the firms. It is a hefty paragraph:

> There's a proposed system called "cap and trade" that some say would lower the pollution levels that lead to global warming. With cap and trade, the government would issue permits limiting the amount of greenhouse gases companies can put out. Companies that did not use all their permits could sell them to other companies. The idea is that many companies would find ways to put out less greenhouse gases, because that would be cheaper than buying permits. Would you support or oppose this system?

The public is divided on the cap and trade approach to reducing greenhouse gas emissions. Averaging across the polls, a slight majority of 51 percent of respondents said they support cap and trade, while 45 percent said they oppose it. Overall, support is much lower than that for direct regulations on carbon emissions—that is, a cap without allowance trading. Support for a cap and trade approach, moreover, is quite sensitive to the price tag. When the potential cost of electricity is included in the question, support for cap and trade falls below 50 percent. *ABC News* and the *Washington Post*, for example, conducted a series of four polls during the 2009 debate over the Waxman-Markey bill. When they asked about support for cap and trade, 55 percent on average favored the measure and 43 percent opposed it. However, when the surveys asked people whether they would support the measure if it raised electricity prices by $25 per month, only 43 percent on average expressed support and 55 percent said they opposed the measure.

The alternative to adjusting the quantity of carbon emitted is to adjust its price. A carbon tax, however, is even less acceptable to the public than cap and trade. Excise taxes on gasoline or electricity or taxes on energy companies that might be passed on to consumers are highly unpopular. Since 2003, the MIT/Harvard Energy Surveys have asked people whether they would be willing to pay more either on their electricity bills or for a gallon of gasoline to "solve global warming." These surveys ask the questions in various ways, but the results that emerge reach the same overall conclusion.

Public willingness to pay higher fuel prices in order to address this problem is very low. As we discussed in the last chapter, in most years the median amount that the public is willing to pay on his or her electricity bill to solve global warming is $5 per month (or $60 per year), or about 5 percent of the typical monthly bill. About 20 percent of Americans are willing to pay at least $25 more per month on electricity, and only a very small percentage of people are willing to pay $100 per month (which would be about twice as much as the current average bill), even though that is the approximate price differential between fossil fuels and alternative energy sources. Looking at the broader range of questions in table 8.1, the same story emerges repeatedly. Only about a quarter of those asked would support a small or modest carbon tax in order to avert global warming.

It is helpful to put these numbers in the context of other sorts of environmental issues. Using similar questions, researchers have discovered similar or even higher rates of willingness to pay for much more specific and localized environmental problems. Numerous studies have estimated that a majority of Americans are willing to pay $10 per month to improve the cleanliness of drinking water in their area.[16] People living in the Mississippi River Basin, for example, are willing to pay $45 more per year for a 10 percent reduction in harmful agricultural runoff that kills fish and makes the river less usable for humans.[17] Boston area residents are willing to pay $2.50 per month to make the Charles River "swimmable," and that river is not a source of drinking water.[18]

The more important context though is how high the tax would need to be in order to make a difference in consumer behavior or energy production. Electricity excise taxes on the order of $5 or $10 per month would have little effect on which sorts of electricity we use or how much we use. The Kyoto Protocol is estimated to imply at least a 25 percent increase in electricity prices in the United States, or $25 per month.[19] Only about one-quarter of the population supports a tax that high. And a tax that made wind competitive with coal would probably have to be much larger. A straight-up carbon tax, then, appears to be a political nonstarter.

Although the unpopularity of carbon taxes has driven many supporters of market-based approaches to climate legislation to focus on cap and trade, it has led others to think about a carbon tax in the context of fiscal policy more generally. A meaningful carbon tax would raise hundreds of billions of dollars in revenue, which would give governments the flexibility to alter other aspects of the tax system. For instance, a carbon tax could be tied either to reductions in personal income taxes or payroll taxes or to reductions in the federal budget deficit. These types of fiscal offsets raise the degree of public acceptance of carbon taxes dramatically, but it is still a far cry from the level of support that regulation receives.

The MIT/Harvard Energy Surveys examined the possible support for such a "tax swap" or revenue-neutral carbon tax. Beginning in 2008, the study considered possible tax swaps.[20] We designed the survey questions to ask about a significant carbon tax, which would translate into a significant reduction in the typical family's annual income and payroll taxes. First, the survey asked whether people supported a $25 monthly surcharge on electricity and a 50 cents per gallon gasoline tax in order to solve global warming. A tax that large would translate into roughly $75 more per month or $900 more per year for the typical family. The survey, then, asked respondents whether they would support such a tax if it was tied to an equivalent amount in income and payroll tax cuts. When respondents were asked about it in isolation, the energy tax received low levels of support. A strong majority of 56 percent opposed such a tax; 29 percent were unsure; only 15 percent expressed support for the tax. However, when the question was tied to reductions in payroll and income taxes, the numbers nearly reversed. Now, a plurality of 41 percent said they would *support* a carbon tax if it were tied to income and payroll tax cuts; 32 percent said they were unsure; only 27 percent opposed such a tax.

We repeated that same approach in the 2011 survey, again with a $25 per month electricity surcharge and a 50-cents-per-gallon gasoline tax. As in 2008, only 25 percent said they would support such carbon taxes, and 50 percent opposed them. The remaining 25 percent were unsure. However, when the energy tax was tied to a commensurate reduction in income and payroll taxes, a clear majority (58 percent) of respondents said they supported the proposal. When we linked a carbon tax with a promise to use the revenue generated to deficit reduction, support reached about 47 percent, about 20 percentage points above that of support for a stand-alone carbon tax. Yet that means a majority still opposed it.

Linking a carbon tax with an income tax cut or with deficit reduction represents for most people a politically more palatable way to implement carbon taxes. These polling data suggest that a carbon tax can be an integral part of an energy policy aimed at reducing climate emissions, but there is insufficient support for a very large carbon tax to make this tool the centerpiece. Moreover, it is important to reiterate that the American public is much more willing to support an aggressive regulatory approach than an equally aggressive set of carbon taxes, even when those taxes are offset by tax reductions. This is a disappointing conclusion for many of our colleagues concerned with the economics of energy and the environment. It commits the United States to a less efficient climate policy than could be adopted, but it is the politically expedient way forward.[21]

Why?

The popularity of a regulatory carbon cap demands an explanation. Why do people prefer a policy option that most experts think is inferior, or at least economically less efficient? The answer becomes clear when we consider why people supported or opposed each of the policies considered in the MIT/Harvard Energy Surveys. Interestingly, the surveys bear out the very narrative developed by the advocates of AB 32 in their efforts to fend off Proposition 23.

One possible explanation—often voiced by advocates of climate policy—is that the public just doesn't understand. Either Americans are insufficiently concerned about global warming, or they do not see the connection between global warming and policies to reduce greenhouse gas emissions such as carbon taxes and cap and trade. If people, the argument goes, only understood the issues (as experts do), they would express more support for these measures. We will show this is not true. As we saw in the previous chapter, most Americans *are* concerned about global warming. The disconnect occurs when people are asked how much they would pay to solve the problem, and that suggests that expressions of concern, for whatever reason, might not translate into meaningful preferences about public policies.

A simple way to gauge whether people's concerns translate into preferences for appropriate public policies is to measure the correlation between expressions of concern about global warming and support for various policies, such as regulation of carbon emissions, carbon taxes, and cap and trade. We do this with multiple regression analysis using the previously discussed response data from the 2011 MIT/Harvard Energy Survey. More specifically, the dependent variables are people's indication of support or opposition for each of the policy alternatives,[22] which we regress on their concern about global warming, their perceptions of the harms and costs of traditional fuels (coal, natural gas, nuclear, oil, and hydro) and alternative energies (wind and solar),[23] and a host of other individual characteristics along the lines of the analyses earlier in the book.[24] (The full results can be found in table A8.1 in the appendix.)

Based on these regression results, figure 8.1 shows the strength of the relationship between support for a given policy and concern about global warming. The reported value for each policy is the difference in predicted probability of supporting the policy for people concerned about global warming and for those not concerned (more precisely, it is the difference in predicted probability for one standard deviation above and below the

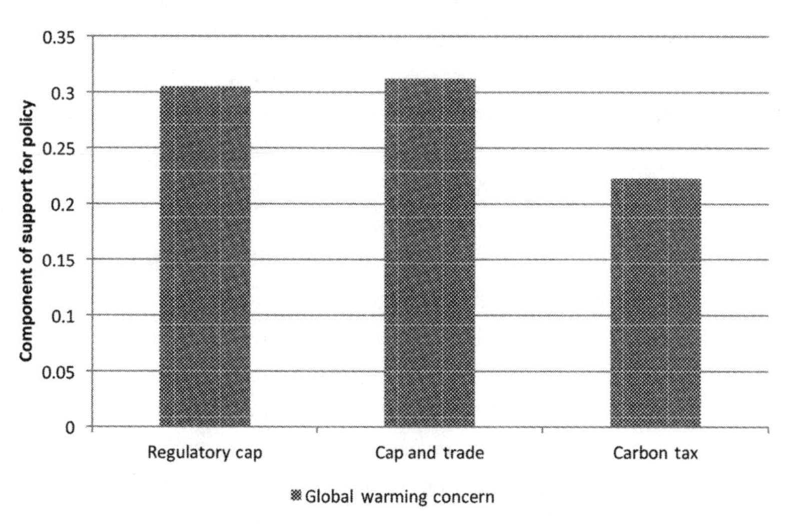

Figure 8.1
Importance of global warming concern on climate policy preferences
Note: Bar graph shows the difference in the predicted probability of support for each policy for a person concerned about global warming compared to a person that is not concerned about global warming. The value is based on the difference in predicted probability for one standard deviation above and below the mean level for each variable, holding the rest of the variables at their mean levels (based on the regression results shown in table A8.1 in the appendix).
Source: 2011 MIT/Harvard Energy Survey.

mean concern level, setting all other factors at their mean levels). In essence, it is the percentage of people who say they support a given policy among those who say they are concerned minus the percentage of people expressing support for a given policy among those who say they are not concerned, holding fixed individuals' income, education, gender, and other factors. Global warming concerns clearly have a very big effect on support for regulations, for cap and trade, and for carbon taxes. The difference is on the order of 25–30 percentage points, which is a very large difference across those who are and those who are not concerned about this issue. That difference is much larger than, say, the effect of education or income or, even, political party. Americans, then, are concerned about global warming, and those concerns, to no small degree, translate into support for various policies. So why then do so many Americans oppose carbon taxes and cap and trade?

Closer consideration of the survey data only deepens the puzzle. Most Americans are concerned or somewhat concerned about global warming *and* these concerns are most strongly correlated with support for cap and trade. Therefore, it seems to stand to reason that cap and trade should be the most popular climate policy (at least among these options). But it is not. Regulation of carbon emissions is by far the most popular policy among the American public and cap and trade is a distant second, garnering the support of only about half of all survey respondents.

Part of the answer to this puzzle lies in the correlation between concern about global warming and support for climate policy. That correlation is reasonably strong, but it is not perfect. A one-to-one correspondence would imply that concern about global warming dictates support for climate policy. The correlation falls far short of perfection, and that, in turn, might reflect either confusion on the part of some people or the presence of other, countervailing considerations in their thinking.[25] The strength or weakness of the correlation, however, is not a sufficient explanation. It cannot explain why regulation (caps alone) are much more popular than cap and trade. Something is missing, and this missing piece, we think, is what has vexed advocates of carbon taxes, cap and trade, and other policies that target greenhouse gas emissions.

The answer snaps into place when we add to the analysis people's opinions about different energy sources used to generate electricity. Just as we did with global warming, we can measure how much support for

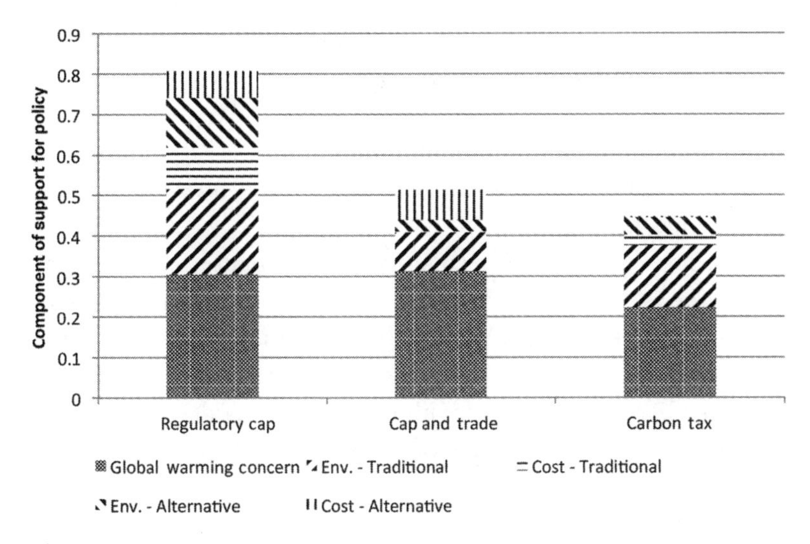

Figure 8.2
Importance of global warming concern and perceptions of environmental and costs of energy sources on climate policy preferences
Note: Bar graph shows the difference in the predicted probability of support for each policy for a person concerned about global warming compared to a person that is not concerned about global warming. The difference is based on the difference in predicted probability for one standard deviation above and below the mean level of concern, holding the rest of the variables at their mean levels (based on regression results shown in table A8.1 in the appendix).
Source: 2011 MIT/Harvard Energy Survey.

each policy is attributable to perceived environmental harms and costs of traditional and alternative fuels. In figure 8.2, we add to figure 8.1 the effects on respondents' support for policies of the respondents' perceptions of the harms and costs associated with different fuels (again, this is based on the difference in predicted probabilities for one standard deviation above and below the mean level of perceived harms and costs). A simple inspection of the graph shows that the additional support for a purely regulatory solution owes to the links that people see between environmental harms and costs of various energy sources and regulation. People associate the environmental and economic aspects of energy with regulatory caps and carbon taxes more strongly than they do with cap and trade. The link that people draw between attributes of energy production and climate policy explains why regulatory caps are much more popular than cap and trade. Simply put, people see a stronger connection between regulation of carbon emissions and other environmental benefits, especially reductions in air and water pollution from reductions in fossil fuel use and increases in alternative fuels. They do not see as strong a link with cap and trade.

The data deserve a closer look since they offer a revealing glimpse into how people view this matter.

First, consider the bar for regulatory caps in figure 8.2. The gray shaded portion is the amount of support for the policy that is attributable to people thinking that global warming is a concern versus not a concern—that's worth about 30 percentage points of support. The portion of the bar that is marked with upward-sloping diagonals shows the effect of environmental harms of traditional fuels. People who see them as relatively harmful express support for a regulatory approach to limit greenhouse gas emissions that is 20 percentage points higher than those who see them as relatively benign for the environment. Cost considerations for these traditional fuels account for another 10 or so percentage points in explaining the overall level of support for regulation. And perceptions of the environmental benefits and economic costs of alternative energy add another 10 and 5 points, respectively. All told, these attributes of energy production added approximately 50 percentage points to levels of support for regulatory limits on greenhouse gases.

Second, contrast that with the accounting of support for cap and trade. Concern about global warming provides a relatively high baseline level of support for this policy, but environmental and economic attributes of energy production added only about 20 percentage points to overall levels of support, compared with 50 percentage points for caps alone. Economic costs and environmental benefits of alternative energy sources showed no statistically significant correlations with support for cap and trade. Concern about environmental harms due to traditional fuels accounts for about 10 points more support for cap and trade, compared with 15–20 points for regulatory caps or for carbon taxes.

Why, then, do more Americans embrace EPA regulation of carbon emissions than support cap and trade? The analyses reflected in figure 8.2 reveal that it is because people see the connection between regulatory caps and local environmental harms, but they do not see such a benefit from cap-and-trade. Limits on greenhouse gas emissions resemble limits on other pollutants, and they would likely lead to reductions in other pollutants. Allowing the EPA to regulate greenhouse gases is an extension of regulatory powers that most Americans already support. And, Americans see that capping carbon emissions would bring immediate benefits from lowering local air and water pollution. Cap and trade, by contrast, was presented to the public in the 2009 debate over the American Clean Energy and Security Act as *climate policy*, not as a tool to fight pollution generally. People do not see a link between cap-and-trade and other benefits. The proponents of the American Clean Energy and Security Act erred in discussing it solely as climate policy. By contrast, the advocates of AB 32 in California saw the connections quite clearly. Our survey data bear out the same lesson: one cannot sell cap and trade or regulatory caps as climate policy alone.

Variation on a Theme: Revenue-Neutral Carbon Taxes

Carbon taxes are even less popular than a cap and trade policy, as was exhibited clearly in the MIT/Harvard Energy Surveys. Yet the failure to pass the 2009 American Clean Energy and Security Act has prompted many policy analysts and advocates to switch their efforts from a push to adopt cap and trade to a quieter campaign supporting carbon taxes. The carbon tax proposals currently under consideration are "revenue-neutral," and are tied to reductions in income or payroll taxes, to reductions in public deficits or debt, or to direct transfer payments to individuals, such as in the British Columbia carbon tax law. Revenue-neutral carbon taxes (defined in various ways) have attracted the support of a politically and ideologically diverse set of people, including Al Gore; the former chair of the Council of Economic Advisors for President George W. Bush, Gregory Mankiw; Nobel Prize–winning economists Paul Krugman and Gary Becker; and the godfather of supply-side economics, Arthur Laffer.[26]

These proposals alter the fundamentals of the policy debate. Regulatory caps, cap and trade, and carbon taxes are, at their core, about global warming and energy production, and that is clearly reflected in the reactions of the American public to such policy proposals. Debates over complicated schemes to trade off one tax against another or to recycle revenues tend to drift into the intricacies of the system of taxes and transfers, rather than energy and the environment. This is a critically important insight to keep in mind as the debate over energy policy begins again and shifts from cap and trade to revenue-neutral carbon taxes. And that fact becomes abundantly clear in the survey data.

We began polling on various ways of structuring carbon taxes in 2006. The idea seemed simple enough. Perhaps it was the case that the intense public opposition to carbon taxes owes to the fact that they are outright consumption taxes with little immediate or evident benefit. Our surveys showed just that. In our 2008 survey, for example, less than 20 percent of Americans said they supported an outright tax of about $25 per month on electricity and $1 a gallon on gasoline (about $75 per month total for the typical household) in order to reduce U.S. greenhouse gas emissions significantly. However, as we discussed earlier, when we asked people if they would support such a tax if it were tied to an equivalent reduction in federal income and payroll taxes, public support rose to 40 percent. That's not a majority, but it is a substantial difference in opinion. We have repeated those same questions in several surveys since, with variations on the exact tax and revenue schemes. The same general pattern emerges: about twenty-five percent of respondents support a significant carbon tax in order to reduce greenhouse gas emissions, and roughly 50 percent of the respondents support such proposals when they are tied in some way to cutting taxes, reducing federal deficits, or transferring income directly back to individuals. Tying carbon taxes to income tax cuts or income transfers boosts support for fuel taxes, but it does not make carbon taxes as popular as regulation.

A deeper concern with such proposals is that they fundamentally change the terms of public debate and understanding of the issues at hand. We examined how people think about various revenue-neutral carbon tax schemes by measuring the degree to which opinions about the harms and costs of energy sources and concern about global warming predict support for each tax proposal. (That analysis is presented in table A8.1 in the appendix.) The difference between that analysis and a similar analysis predicting support for carbon taxes alone is striking. The relation between support for a carbon tax and concern about global warming is cut in half (or more) when the carbon tax is tied to reductions in other taxes or deficits. Concern about environmental harms from different energy sources is only weakly related to support for revenue-neutral carbon taxes. And expressions of support or opposition to the tax are more sensitive to fuel prices. In short, people think about complex schemes for revenue-neutral carbon taxes more in terms of fuel prices and tax reform and less in terms of environmental harms, even global warming. This amounts to selling energy and environmental policy as tax policy. That might appear to be a clever political move, or it might backfire.

Our analysis of public attitudes about climate policy provides factual support for what one might have intuited. People who care about global warming are more likely to support policy aimed at reducing greenhouse gases. That relationship, however, is only part of the story. It takes more than just concern about global warming to win support of a majority of the public for climate policy. The political fate of various climate policies depends primarily on the other half of the equation—how the public thinks about energy.

The political error of many experts and advocates for climate policy is that they have presented the case for their proposals and ideas too narrowly or, worse, have developed regulatory mechanisms that are too specifically focused on greenhouse gases. A regulatory approach to climate change is less efficient, according to our colleagues in economics; one can more efficiently take the carbon out of the economy by setting a price on carbon through a cap and trade system or a tax. That approach targets carbon emissions more effectively than the blunt hammer of the EPA setting a limit on carbon. But EPA regulations are politically the most palatable option today.

The reason Americans support regulation over carbon taxes and cap and trade is that they connect limits on greenhouse gas emissions to other energy and environmental benefits that such policies would have. That is, Americans see a regulatory approach to climate change not just as a climate policy, but as an energy and environment policy generally. Allowing the EPA to set limits would not only reduce greenhouse gas emissions but also help satisfy unmet public demand for a cleaner environment by reducing air and water pollution from fossil fuels and increasing our reliance on alternative sources of energy. And they may not be wrong about that. Cap and trade might lead to reductions in greenhouse gas emissions in industries that operate far from population centers but do little to reduce smog and other co-pollutants of carbon dioxide emissions.

Co-pollutants such as smog in urban areas may be more expensive to reduce than, say, methane emissions from mines, but they may create greater health benefits for people today. People see more clearly the links between regulation and reducing pollution in their own backyards, and those benefits appeal powerfully to the American public.

Public attitudes about climate policy, then, are an extension of how people think about energy and the environment. That fact was certainly understood by the opponents to Proposition 23. By casting cap and trade as a policy to address traditional pollution problems and to fight oil companies, the opponents to Proposition 23 were able to defend what was fundamentally a climate policy. And that insight provides an opening for building majority support for policies needed to reduce greenhouse gas emissions.

9

A Way Forward

We are at the beginning. It is the beginning of a decades-long transformation of America's energy system. It is the beginning of a century-defining struggle with climate change. America will lead the way. We have no choice, as the world's largest economy, one of the world's largest producers of oil, natural gas, and coal, and the second-largest emitter of greenhouse gases. The push is on to make our energy system cleaner without destroying our economy or making energy prohibitively expensive for large segments of the population.

The question is, how do we get there? That is not clear, but the public's role, either as a constructive force or as an obstacle to needed policy, is undeniable. In this regard, the story of energy is probably not much different from other transformations of public policy in U.S. history. As generations of political science research have shown, the policies pursued by the government over the long run track with the ups and downs of public opinion about the significant issues facing the country.[1] Public consensus, once it emerges, is a very powerful force.

Our reading of U.S. public opinion about energy over the past decade points to a significant opportunity to address the climate problem, even without a consensus on the climate issue. Don't just treat climate as climate. There is considerable support for the U.S. Environmental Protection Agency to set regulatory caps on carbon emissions, and for other policies that would indirectly cap emissions such as renewable portfolio standards. Support for these policies is even greater when discussed in the context of local environmental harms, as exemplified in the fight over California's Global Warming Solutions Act.

But the public will support other approaches too. In fact, the nation might even ignore the climate issue and make progress on the problem.

The strategy is to target aggressively the co-pollutants of carbon found in many fuels, such as particulates, sulfur, and mercury. These present immediate and localized risks to health, and the social benefits of reducing such pollutants (by reducing the use of the fuels with the highest concentrations) would have a substantial benefit to society today. Most Americans support legislative and regulatory actions to address these pollutants, even lowering the use of fuels such as coal and oil. EPA regulation of these pollutants, using authority it already has, would change the cost calculations of industry, making alternative fuels more competitive and reducing use of the more polluting fuels and products produced with these fuels. The public today cares much more about local environmental harms from the energy system precisely because the social cost of those harms far exceeds the present value to society of the cost of climate change in decades to come. The advocates of climate policy have been so focused on this problem unto itself that they often miss this important insight. The climate question should be viewed from a different vantage point, from the perspective of energy and the environment more broadly. On questions of immediate environmental regulations, public opinion approaches a consensus, or at least a majority large enough to push government to act.

Over the coming decades, the push of the American voter and the pull of the American consumer will shape the decisions that governments and firms make about the development of energy. Public opinion has already started to drive state governments toward an alternative energy future. Texas made a dramatic investment in wind energy a decade ago following extensive study using focus groups and deliberative polls to gauge what the average Texan wanted. California, through both its legislature and the ballot, adopted a cap and trade system. State legislatures in Connecticut, Delaware, Maine, Maryland, Massachusetts, New Hampshire, New Jersey, New York, Rhode Island, and Vermont created the Regional Greenhouse Gas Initiative. And, thirty state legislatures and electorates have adopted some form of a Renewable Portfolio Standard. The public will also constrain industry and government from going down some avenues. Most likely, continued unease with the safety and waste of nuclear power will prevent a massive deployment of this technology. That is not a prediction based on technology trends or innovations in engineering laboratories, but on the persistent concerns that Americans harbor about nuclear power safety and the long-term storage of spent nuclear waste.

The disparate trends and reactions to public opinion about energy among the states indicate that industry and government will continue to listen to what people want in the energy domain. But it is not enough to read the latest poll numbers on this or that issue. Policy makers, investors, and engineers need to listen carefully to what people want. That is where the demand for specific sorts of power lies; that is how public support on this issue is won or lost. For decades, however, public opinion research on energy has been highly fragmented and episodic. Survey researchers have typically asked about the latest sensational event or about one power source (most notably, nuclear power) in isolation from the other choices. That is like trying to figure out what the patient wants by looking only at a superficial symptom or by taking one reading of his or her vital signs.

Listening to what the American public wants involves thinking and asking about the entire energy system. It is not meaningful to ask about nuclear power in isolation, because the energy policy and energy business questions are ultimately not "nuclear power yes or no," but "nuclear power compared to what." In short, to understand public opinion about energy in a meaningful way requires that pollsters and analysts think systematically about energy. What are the main sources of energy, and what are the public's attitudes toward each one? What are the key attributes of each energy source, how do people view each energy source according to those attributes, and how important are each of the attributes in explaining what people want?

Public opinion surveys show unambiguously where the public wants the energy industry to go. There is strong and clear support for increasing use of solar and wind power. A majority of Americans support expanding the use of these alternative energy sources, at least in principle. There is also a clear desire to move away from coal and oil, at least as we use them today. An outright majority says it wants to reduce the use of coal and oil, and the desire to reduce the use of oil has tracked steadily downward over the past decade.

On the whole, however, public opinion researchers to date have done a poor job gauging public attitudes about energy in a way that can guide the development of a comprehensive and sustainable energy policy.

Typically, survey work about energy has been limited to a particular controversy or a narrow segment of the energy market, such as yesterday's spike in gas prices or an oil spill off the coast. When that news fades, so too does the polling, and it never really adds up to much. There are notable, recent exceptions to this statement, such as James Fishkin's deliberative polling in Texas.

Our goal has been to fill the gap in understanding, at least somewhat, by developing a new, systematic approach to public attitudes about energy. We have presented a Consumer Model of what people want from energy and energy policy. In that model, an individual's opinion about the use of various fuels and the direction of public policy depends on two key factors or attributes of energy—the economic cost (the private good) and the environmental cost (the public good). We have demonstrated throughout this book that these two attributes, economic and environmental costs, shape public opinion about what sources of energy people want to be brought to market and about what energy and environmental policies the government ought to pursue.

Consumers and Voters

The Consumer Model helps us understand what people want and why. Do most Americans think that we should use more of some and less of other sources of power? Do most Americans want to use less of all sources of energy? How important are prices and harms in people's thinking? And how do those compare with partisanship, ideology, income, region, and other factors? What would it take to bring Americans to support an aggressive carbon tax or cap and trade?

We have tracked these questions over the course of the past decade to see what the public wants the energy sector to look like and how the public's preferences have evolved. A stable picture of both what the public wants and how the public thinks emerges. In terms of what people want, the story is quite simple. Very few Americans are strictly conservationist, desiring to use less of all forms of energy. Most Americans want to increase substantially our use of wind and solar power. Moreover, most people do not want the United States to use less of all traditional fuel sources. Most would reduce our reliance on coal and oil, but increase or keep the same our use of natural gas. And, in fact, that is exactly what has happened over the past decade. Natural gas has surged, in part due to technological innovations that have reduced prices and in part due to consumers' willingness—even at the expense of retrofitting their home energy systems—to make use of a fuel that they see as cheaper and cleaner than the alternative.

More surprising are our findings on how people think about energy. Energy is not a highly politicized issue, pitting Democrats against Republicans or liberals against conservatives or rich against poor or Northerners against Southerners. People's attitudes about what energy we use and what policies the government puts in place depend primarily on understanding the prices and environmental harms associated with specific energy sources. And people tend to think the same way about all fuels. That is, people respond to price variations in approximately the same way for all fuels. They are not more price sensitive when it comes to, say, nuclear power than they are for wind or gas. People also tend to weigh environmental factors the same for all energy sources. (The exception may be nuclear power where the particular concerns of waste storage and safety seem to have unusually strong affects on people's thinking.) In the public's mind, then, energy is energy—no proper names are required.

Where the fuels differ is in their actual attributes. Some fuels are cleaner than others, and people, on average, have the right perception of the relative harmfulness of various fuels. Some fuels are cheaper than others, and people, on average, have the costs right, with one caveat. People think that providing a large amount of our electricity from solar and wind power would be very cheap, in absolute terms and relative to coal, natural gas, and nuclear. The reason that the public is bullish on alternative energy is that a large majority rightly sees these fuels as clean relative to traditional fuels and highly values environmental protection; however, the public incorrectly views these fuels as relatively inexpensive. When people learn the true prices, support for these fuels falls.

A very important political implication follows. Our colleagues in economics and engineering and many advocates often blame public ignorance for the state of U.S. energy policy. Many reports call for greater public education on energy, implying that things would be different if only people knew the truth. That is never a bad thing, but it might have the opposite effect to what the advocates want. If there were an extensive public education campaign on the costs and environmental impacts of various energy sources, our data indicate, there would be relatively little movement on many of the basic indicators concerning public support or opposition to various fuels. The reason is that, on average, the public's perceptions of the energy area are about right. To the extent that the public systematically errs, it is in the belief that alternative fuels are very cheap. Educating people about the true costs of heavy reliance on electricity from solar and wind power might thus make people less supportive of the very power sources that many advocates want increased.

Our results also explain an important puzzle about Americans' preferences about public policy. Economists and other experts have long been troubled by the fact that Americans—in public opinion surveys and at the polls—often seem to prefer less efficient energy policies to more efficient ones. Voters choose regulation over cap and trade, even though cap and trade holds the possibility of reducing emissions with less drag on the economy. And this is true even when concern about the issue at stake (global climate change, in our case) is more strongly correlated with the efficient policy. The Consumer Model helps us understand why that is the case. People draw the connection between energy prices and harms and regulation, but those connections are lacking when it comes to complicated or unfamiliar policies such as cap and trade.

The Consumer Model, then, helps deepen our understanding of public opinion. It provides a broader framework within which to interpret the myriad and fragmented survey findings and polls that often seem at first blush like just many scattered tea leaves.

Linkages

The real power of thinking systematically about public opinion and energy is that doing so opens up new opportunities. Americans know what they want, but it is up to our leaders in industry and government to see where we can go as a nation and to present us with the choices that offer a real improvement for our society. In that respect, the analysis that we have presented reveals five key linkages between different aspects of energy, environment, and public policy. They inform how we think about energy and public opinion and, ultimately, point a way forward.

Link 1: Energy and the Environment. The private energy market in the United States is subject to relatively little regulation compared to other countries. The main regulatory framework that shapes the development of energy is the web of federal and state environmental laws and regulations. As Michael Graetz points out, we do not have an energy policy; we have an environmental policy that serves as our energy policy.[2] That is an inefficient and confusing way to achieve the sort of energy system that the American public wants, but it reflects an essential fact: energy and the environment are inextricably linked.

Technology creates the linkage between energy and the environment. Fossil fuels, and before them wood, have driven the economic and industrial development of the United States throughout our history. The land is rich in these natural resources, and they have powered the American economic engine. The unfettered use of these fuels unfortunately comes with high social costs. The Clean Air Act of 1970, the Clean Water Act of 1972, and other environmental laws provided the architecture for gauging the social costs of energy production and other activities and forced consumers and industry to reduce pollution. These rules brought higher prices on cars, motor vehicle fuel, electricity, and the like, but they also yielded society-wide gains through the reduction of lung diseases, cancers, and other ills. Efforts to rein in the social costs of pollution have thus created an indirect sort of energy policy. And as long as we maintain a fossil fuel–based economy, environmental laws will shape our energy use and serve as a form of energy policy.

Today, the Clean Air Act is being extended from its historical focus on conventional pollutants such as ozone, sulfur dioxide, nitrogen oxides, and particulate matter, and hazardous air pollutants such as mercury and benzene. In 2007, the U.S. Supreme Court ruled that the EPA has the authority to regulate carbon, methane, and other greenhouse gases as pollutants. The EPA has begun the process of making the appropriate rules, and those rules will shape the energy sector in the United States.

Link 2: Fuels and Their Costs and Harms. As we have shown throughout this book, public perceptions of and attitudes about coal, oil, natural gas, nuclear power, wind, solar and other sources of energy are shaped by two key attributes—economic costs and environmental harms (or social costs). People think about energy as consumers, not as partisans. They want energy that is clean and cheap, and that is a driving force in the consumer market for energy.

Americans strongly value both attributes. To the extent that people experience or believe that a fuel is cheap or clean, they will support it. Any fuel that is seen to be cheaper and cleaner than alternatives is strongly preferred by the public. These are positive links—we want more of both.

Link 3: Economic Costs versus Environmental Harms. Not all links are positive. Such is the case with the economic costs and environmental harms associated with various fuels. Because of the state of the existing technologies and the nature of the fuels, there is a real trade-off between the economic and environmental attributes of energy sources. That is, currently no fuel exists that is both cheap and clean. Coal is very abundant in the United States and relatively inexpensive, but it is quite dirty to mine and burn. Wind is intermittent and relatively expensive, but it produces no air or water pollution. Natural gas lies in between these poles. It is cleaner than coal but does emit some pollutants. Until recent advancements in hydraulic fracturing and horizontal drilling, natural gas has been more expensive than coal but cheaper than wind. There is a very good reason for this trade-off. Any fuel that is both very damaging to the environment and expensive gets driven out of the market. Today, the main choices are among coal, natural gas, nuclear power, oil, and wind; and there is a glimmer of hope for solar power. None currently dominates. Which of these and other fuels industry and consumers favor depends on how the marketplace will balance economic and social costs.

The key to understanding public opinion about energy is *how people make the trade-off* between the economic and social costs of energy production and use. How sensitive are people to electricity prices? How sensitive are their attitudes to local pollution harms or to global environmental risks? The central thrust of our research has been to gauge in which direction most people lean. Do they prefer a modest increase in economic cost to decrease social costs or to reduce emissions (even if they do not affect them immediately), or do they prefer to keep energy prices low even at the risk of environmental damage and other social costs? Most Americans tilt in the direction of the environment. The public values both attributes, but our survey data reveal that most people's support for energy sources is more sensitive to perceptions of environmental harms than to economic costs.

There is a very good reason for this. Private energy firms and the marketplace are quite good at delivering reliable, low-cost energy. If one fuel becomes more expensive, firms will substitute it for another source. Energy firms and the marketplace are less willing, however, to remedy the environmental damage produced by the production process or use of a given fuel. There's no margin in environmental cleanup. As a result, there is pent-up demand among consumers and voters for the reduction of social costs associated with energy production. Relatively small increases in energy prices have the potential to make substantial improvements in health, but these are not presently reflected in our energy prices.

Link 4: Local Environment and the Global Environment. Much of the pollution that results from generating electricity is local in nature. Although air and water pollutants travel to neighboring states and sometimes other countries (think acid rain), their origins and effects are fairly localized. Carbon emissions have global environmental consequences owing to their concentration in the Earth's atmosphere, even though these emissions have mild or no local pollution effects. Even still, local and global environmental harms are positively linked to one another because of the nature of these fuels and the technology today. Coal, oil, and natural gas emit chemicals and particulates that produce many of the most severe local environmental harms, and these fuels produce substantial greenhouse gas emissions when burned. Carbon, then, is a co-pollutant with the particulates, sulfur dioxide, nitrogen oxides, mercury, and other chemical emissions that come with burning large amounts of coal, oil, and natural gas. Any regulation or price on carbon will affect the emission of other pollutants that reside in a given fuel. Likewise, any regulation on mercury, sulfur dioxide, methane, and the like will affect the cost of using coal or oil or natural gas, and thus carbon emissions.

This linkage appears in public perceptions, but subtly so. Concern about global warming is correlated with higher support for alternative energy sources, and in more recent years lower support for coal and oil. Increasing the salience of global warming or of local harms associated with coal and oil pushes people further in the direction of support for alternative fuels and away from oil and coal. For example, when people are told both the social costs of using coal and the social costs of global warming, we saw substantial increases in support for wind, solar, and nuclear power, and declines in support for coal and oil. This increase in support for low-carbon fuels exceeded what one would expect from either local harms or global risks alone.

That said, local environmental harms seem to be an easier sell. Perceptions of air and water pollution associated with a fuel have a much stronger effect on support or opposition for that fuel than do concerns about global warming. At the beginning of the 2000s, concern about global warming was almost entirely uncorrelated with support or opposition to coal, oil, natural gas, and wind power, and only somewhat correlated with support for solar power. The correlation between concern about global warming and support for alternative fuels and opposition to fossil fuels has grown somewhat over the course of the past ten years, but the correlation between perceived environmental harms associated with each fuel and support for expansion of that fuel is much stronger.

Link 5: Climate Policy and Energy Attributes. Americans view climate policy through the lenses of energy use and the costs and harms associated with different types of fuels. Americans already see that climate policies, such as regulatory caps, cap and trade, and carbon taxes, will affect economic costs and environmental harms associated with various fuels. People's perceptions of the harms and costs of traditional and alternative fuels are significant factors in explaining support or opposition to climate policies. The effect of such local environmental harms on support for climate policies is as substantial as people's concerns about global warming itself.

The degree to which people see climate policy through the lenses of local environmental harms and economic costs associated with different energy sources varies across policies. In the case of caps on carbon and carbon taxes, perceptions of local environmental harms are just as important as concern about global warming. Cap and trade is exceptional. After the debate over the American Clean Energy and Security Act of 2009 (the Waxman-Markey bill), Americans viewed cap and trade as almost entirely about global warming. Those concerned about global warming supported the legislation; those who would choose a more cautious approach or who are skeptical about global warming opposed cap and trade. Concerns about local harms or economic costs associated with different energy sources were, at best, secondary.

This difference reflects, we think, the different dialogue about cap and trade, and the more narrowly defined nature of the policy instrument. Carbon taxes and EPA regulation of greenhouse gases are debated in the context of energy taxes and environmental regulations generally. Giving the EPA the power to limit carbon emissions would only further its authority to regulate pollution generally. Taxing gasoline and electricity would change consumption behavior and reduce many pollutants, not just carbon emissions. Cap and trade was presented and debated as about carbon emissions, and carbon emissions alone. Proponents did not present any side benefits or discuss this as a general approach to pollution. This may have been a rhetorical mistake, or it may be a true reflection of what such a mechanism would do—price and reduce carbon, leaving other pollutants roughly the same. The proponents of the bill were able to rally the support of those who are deeply concerned about global warming and to swing some others to their cause. But that did not create a large enough majority of the American public to carry such legislation through both the House *and* the Senate.

A Way Forward

The political error of advocates of climate policy so far has been to think of and discuss greenhouse gases and climate change in isolation of other aspects of energy production. The public, however, does not think that way. Most Americans view climate change through the lens of energy, especially the economic and social costs associated with energy production and consumption. Yes, concern about global warming is a major predictor of who supports government action on climate change, and increasing concern about global warming will increase support for polices designed to reduce greenhouse gas emissions. But equally strong predictors are perceptions of the local environmental harms and economic costs of fossil fuels and alternative energy sources. Supporters of the American Clean Energy and Security Act of 2009, of which cap and trade was a central feature, focused their arguments on the social benefits associated with addressing the challenge of climate change, while the detractors smartly emphasized the economic cost associated with such laws. The link between reductions in local environmental harms and climate policies was lost in the conversation.

The campaign to defeat Proposition 23 is the exception that proves the rule. Opponents of the proposition, as we discussed in chapter 8, sought to protect AB 32, which required the creation of a cap and trade system to impose a limit on California's greenhouse gas emissions. They rarely even mentioned global warming. Instead, their defense of the bill rested on distrust of oil and electricity companies and the environmental and health damages associated with coal and oil. Those crafting climate policy and those advocating for it should take to heart the lesson of the Proposition 23 campaign.

The public opinion data summarized in this book shows that there is a substantial policy opening, but not through the obvious approach of enacting a comprehensive climate law. Large numbers of Americans support EPA regulation to limit carbon emissions. The polling data offer further insight into the politics of how to proceed. First, regulate the co-pollutants of carbon and other greenhouse gases. Oil, natural gas, and especially coal are the primary sources of human made greenhouse gases, but they are also the source of other forms of pollution that the EPA already has the authority to regulate under existing environmental statutes.

Michael Greenstone and Adam Looney propose that domestic pollution regulations be designed to reflect the true social costs of energy production and use. Air and water pollution from energy use have external costs on society in terms of poorer public health. Smog causes child and adult asthma, mercury causes birth defects, lead causes retardation, and so on. Those costs can be monetized in terms of health care treatment and loss of worker productivity. We pay these costs throughout our economy, but they are not reflected in the price of energy, so energy firms have no economic incentive to eliminate them. Greenstone and Looney estimate that the true social cost of fossil fuel energy (ignoring global warming) is approximately 75 percent more than we currently pay for energy. In other words, if the price of energy reflected the true social cost, we would pay much more for electricity from coal, oil, and natural gas, perhaps as much as 75 percent more.[3]

Regulations that capture the true social costs of energy production would drive up the price of electricity from fossil fuels, and would without question hit coal particularly hard. The price of electricity from existing coal-fired power plants would nearly double, from 3.5 cents per kilowatt per hour to 6.5 cents per kilowatt per hour, and the price of electricity from new coal power plants would increase from 6.25 cents to 9.75 cents per kilowatt per hour. At those prices electricity from new wind and nuclear plants would be as cheap or cheaper than electricity from new coal plants, and the price differential between new wind and existing coal plants would be reduced greatly.[4] At that price point, a modest carbon tax becomes feasible, and a modest carbon tax on top of these regulations would make coal too expensive.[5] Such a regulatory goal would greatly accelerate the adoption of technologies needed to reduce our carbon emissions.

This approach need not preclude the use of coal in the future. As the MIT study "The Future of Coal" concludes, higher prices on the use of pulverized coal- fired power plants (the most widely used technology today) would make competitive other coal technologies, such as integrated gasification combined cycle (IGCC) with carbon capture and sequestration.[6] The most likely effect, though, would be to give natural gas a considerable advantage in the market and to make nuclear power and wind competitive with coal.

Greenstone and Looney's proposal rests on an important assumption—that the public actually wants tighter restrictions on conventional pollutants. Our analysis in chapters 5, 6, and 8 reveals that is indeed what the public wants, and they view cleaner energy sources such as wind and solar power as a means to that end. The public rightly perceives that wind and solar offer much cleaner ways of generating electricity than coal, oil, and natural gas. In addition, the main factor driving public support for specific energy sources is to avoid the health harms that come with their use. A one-unit difference in perceived harmfulness of an energy source has about two times as large an effect on support for that energy source as does the perceived price or cost of that energy source. To put this matter another way, the popularity of solar and wind power reflect unmet public demand for cleaner air and water, and the public supports stricter environmental regulations as a way to meet that demand. The insight offered by Greenstone and Looney is that such regulations also take us a long way toward making alternative energy cost competitive with coal, oil, and natural gas. Strengthening traditional environmental regulations serves the same ends as enacting new regulations on carbon emissions.

Regulating co-pollutants of carbon, such as sulfur, particulate matter, and mercury, is a first step. Such regulation would not face the same degree of public opposition as carbon taxes or cap and trade because a large majority of the public currently embraces more aggressive enforcement and expansion of conventional environmental regulations. This approach would lead to the entry of alternative technologies (such as IGCC coal) and other fuels because it would eliminate much of the price advantage that traditional pulverized coal enjoys. Regulations of co-pollutants (such as mercury) would further reduce the magnitude of any other policies required to keep carbon emissions low. Stronger air and water pollution regulations associated with coal-fired power generation would make natural gas competitive with coal, and would eliminate most of the price differential between coal and nuclear power or wind power. A small carbon tax or modest cap and trade system could eliminate much of the remaining difference.

Perhaps most important, regulating coal, oil, and other fossil fuels so that their production reflects social costs associated with pollution would change the long-run trajectory of the American energy sector. Such a policy would make coal prices reflect the true social costs as well as the economic costs. This would, in turn, create a strong economic incentive to develop new technologies for future power generation that would have less pollution impact.

A World without Trade-Offs

Ultimately, public opinion will shape not just what choices are made today, but what options are available to future generations. On this score, we are very optimistic. The history of energy in the United States (even in the absence of a coherent public policy) leads us to hope.

Much of this book has considered the trade-off between energy sources that are cheap and energy sources that are clean. That trade-off lies at the heart of the energy problem facing the United States. The public values two distinct attributes in the energy needed for a robust economy and a healthy society, but those attributes in practice function at cross-purposes. At least in the near term, there are no energy sources that are both low price and low harm.

But the public wants a different energy future. It wants significant expansion of alternative energy not because people like the sun and the wind, but because they see those energy sources as much less polluting than fossil fuels and believe, often wrongly, that those fuels are relatively inexpensive. They want less coal and oil, primarily because those fuels are perceived as very harmful to the environment and, in the case of oil, increasingly expensive. It is the objectives or attributes that people desire, not the fuels themselves.

People—acting as consumers and voters—constantly express their demand for energy that is cheaper and cleaner. At times it feels that the American public is asking for the impossible, and in the short run that is probably true. Over the long run, however, the market working in tandem with the government has delivered the impossible: reliable energy provided at steadily lower prices and with less environmental damage. Over the long run, the amount of income generated in the economy per unit of energy used has risen. Over the long run, energy prices in the United States have fallen steadily. Over the long run, the amount of pollution per unit of energy used has dropped precipitously. These changes have come about through the development of new technologies for more efficient production and use of existing fuels, through the development of new technologies to exploit as yet untapped sources of energy, through the design of more efficient energy delivery systems, and through government regulation of pollution, especially the implementation of the Clean Air Act and the Clean Water Act.

Today, we have much cleaner and much less expensive energy than at any time in our past. Engineers' innovations, energy companies' investments, and government actions have made that possible. Their actions ultimately came about in response to strong demand from consumers for more and better energy systems, and the American marketplace, working through industry, government, and research organizations, has proved remarkably robust at meeting what the public wants. The climate change problem undoubtedly makes the challenge more difficult moving forward. But the public's demand for cheap and clean energy has driven America's energy past, and it will drive America's energy future.

Appendix

Table A3.1
MIT/Harvard Energy Surveys

Year	Survey firm	Sponsor
2002	Knowledge Networks	Massachusetts Institute of Technology
2003	Knowledge Networks	Massachusetts Institute of Technology
2006	Knowledge Networks	Massachusetts Institute of Technology
2007	Knowledge Networks	Massachusetts Institute of Technology
2008	Knowledge Networks	Massachusetts Institute of Technology
2009	Knowledge Networks	Harvard University
2010	YouGov/Polimetrix	Harvard University
2011	YouGov/Polimetrix	Harvard University
2013	Knowledge Networks	Georgetown University

Table A4.1
Perceived costs of fuels across demographic groups; 2002, 2007, 2008, 2011 surveys pooled

	DV: Perception of cost (1 = Very expensive, 5 = Very cheap)						
	Coal	Natural gas	Oil	Nuclear	Hydro	Solar	Wind
Age 18–29	0.01	0.00	0.12**	-0.11*	-0.26**	-0.31**	-0.25**
	(0.045)	(0.038)	(0.039)	(0.047)	(0.043)	(0.052)	(0.050)
Age 60+	0.08*	0.10**	0.08**	0.25**	0.22**	0.23**	0.20**
	(0.037)	(0.031)	(0.032)	(0.038)	(0.035)	(0.043)	(0.041)
Minority	-0.16**	-0.11**	-0.02	-0.14**	-0.14**	-0.10*	-0.06
	(0.039)	(0.033)	(0.033)	(0.040)	(0.037)	(0.045)	(0.043)
Female	-0.16**	-0.22**	-0.12**	-0.34**	-0.27**	0.08*	0.06
	(0.031)	(0.027)	(0.027)	(0.033)	(0.030)	(0.037)	(0.035)
Education	0.08**	0.07**	0.05**	0.09**	0.02	-0.11**	-0.12**
	(0.014)	(0.012)	(0.012)	(0.015)	(0.014)	(0.017)	(0.016)
Income	0.13**	0.12**	0.06**	0.09**	0.05**	-0.02	-0.04*
	(0.017)	(0.014)	(0.015)	(0.018)	(0.016)	(0.020)	(0.019)
Democrat	-0.03	-0.11**	-0.12**	-0.17**	-0.10*	0.09	0.11*
	(0.043)	(0.037)	(0.037)	(0.045)	(0.041)	(0.051)	(0.048)
Republican	0.12**	0.10**	0.16**	0.20**	0.07	-0.06	-0.10*
	(0.045)	(0.038)	(0.039)	(0.047)	(0.043)	(0.053)	(0.050)
Northeast	-0.02	-0.04	-0.05	0.04	0.06	-0.05	-0.02
	(0.048)	(0.041)	(0.042)	(0.050)	(0.046)	(0.056)	(0.054)
South	-0.10*	-0.00	-0.07	-0.02	-0.06	-0.05	-0.01
	(0.041)	(0.035)	(0.036)	(0.043)	(0.039)	(0.048)	(0.046)
West	-0.12*	0.17**	0.05	0.04	-0.06	-0.11*	-0.11*
	(0.046)	(0.039)	(0.040)	(0.048)	(0.044)	(0.054)	(0.051)
2007 survey	0.10*	-0.13**	-0.15**	0.19**	0.21**	0.03	-0.09
	(0.046)	(0.039)	(0.040)	(0.048)	(0.044)	(0.054)	(0.051)
2008 survey	0.01	-0.07	-0.20**	0.33**	0.22**	0.13*	-0.04
	(0.046)	(0.039)	(0.040)	(0.048)	(0.044)	(0.054)	(0.051)
2011 survey	0.41**	0.52**	0.22**	0.56**	0.48**	-0.13*	-0.41**
	(0.044)	(0.037)	(0.038)	(0.046)	(0.042)	(0.052)	(0.049)
Constant	2.80**	2.54**	2.09**	1.99**	3.17**	3.67**	4.11**
	(0.064)	(0.054)	(0.056)	(0.067)	(0.061)	(0.075)	(0.072)
N	5,025	5,087	5,116	4,921	4,947	5,160	5,099
R-square	0.080	0.153	0.072	0.128	0.089	0.044	0.065

Note: Cells contain OLS regression coefficients, with standard errors in parentheses. Significance: $^{**}p < 0.01$, $^{*}p < 0.05$

Table A4.2
Perceived environmental harms of fuels across demographic groups; 2002, 2007, 2008, 2011 surveys pooled

	DV: Perception of harm (1 = Very harmful, 5 = Not harmful at all)						
	Coal	Natural gas	Oil	Nuclear	Hydro	Solar	Wind
Age 18–29	-0.08	-0.30**	-0.18**	-0.14**	-0.10*	-0.19**	-0.07*
	(0.045)	(0.041)	(0.044)	(0.050)	(0.043)	(0.029)	(0.029)
Age 60+	0.11**	0.24**	0.22**	0.49**	0.20**	0.05	0.04
	(0.037)	(0.034)	(0.036)	(0.042)	(0.035)	(0.024)	(0.024)
Minority	0.04	-0.10**	0.10**	-0.35**	-0.22**	-0.25**	-0.19**
	(0.040)	(0.036)	(0.038)	(0.044)	(0.037)	(0.025)	(0.025)
Female	-0.02	-0.16**	-0.01	-0.57**	-0.11**	-0.03	-0.01
	(0.032)	(0.029)	(0.031)	(0.035)	(0.030)	(0.020)	(0.020)
Education	-0.10**	-0.01	-0.10**	0.14**	-0.02	0.04**	0.02*
	(0.015)	(0.013)	(0.014)	(0.016)	(0.014)	(0.009)	(0.009)
Income	-0.03	0.07**	0.01	0.15**	0.05**	0.06**	0.03*
	(0.017)	(0.016)	(0.017)	(0.019)	(0.016)	(0.011)	(0.011)
Democrat	-0.28**	-0.13**	-0.23**	-0.19**	-0.13**	-0.04	-0.02
	(0.044)	(0.040)	(0.042)	(0.049)	(0.041)	(0.028)	(0.028)
Republican	0.38**	0.32**	0.43**	0.48**	0.22**	-0.05	-0.06
	(0.046)	(0.042)	(0.044)	(0.051)	(0.043)	(0.029)	(0.029)
Northeast	-0.17**	-0.03	-0.01	-0.10	-0.06	-0.02	-0.01
	(0.049)	(0.045)	(0.047)	(0.055)	(0.046)	(0.031)	(0.031)
South	-0.09*	-0.03	-0.09*	0.03	-0.04	-0.01	0.00
	(0.042)	(0.038)	(0.040)	(0.047)	(0.040)	(0.027)	(0.027)
West	-0.22**	0.08	-0.21**	-0.01	-0.12**	0.02	0.02
	(0.047)	(0.043)	(0.045)	(0.052)	(0.045)	(0.030)	(0.030)
2007 survey	0.10*	0.08*	0.10*	0.30**	0.23**	0.16**	0.03
	(0.047)	(0.043)	(0.045)	(0.052)	(0.044)	(0.030)	(0.030)
2008 survey	0.10*	0.22**	0.18**	0.44**	0.23**	0.20**	0.11**
	(0.047)	(0.043)	(0.045)	(0.052)	(0.044)	(0.030)	(0.030)
2011 survey	0.53**	0.55**	0.58**	0.47**	0.27**	0.13**	-0.02
	(0.045)	(0.041)	(0.044)	(0.050)	(0.043)	(0.029)	(0.029)
Constant	2.50**	3.35**	2.61**	2.04**	4.00**	4.57**	4.67**
	(0.066)	(0.060)	(0.063)	(0.073)	(0.062)	(0.042)	(0.042)
N	5,358	5,296	5,368	5,292	5,234	5,438	5,427
R-square	0.097	0.120	0.114	0.214	0.061	0.059	0.022

Note: Cells contain OLS regression coefficients, with standard errors in parentheses. Significance: $^{**}p < 0.01$, $^{*}p < 0.05$

Table A4.3
Correlations among perceptions of energy source cost and harm

	Perceptions of cost						
	Coal	Natural gas	Oil	Nuclear	Hydro	Solar	Wind
Coal	1.0						
Natural gas	0.66	1.0					
Oil	0.44	0.49	1.0				
Nuclear	0.44	0.52	0.36	1.0			
Hydro	0.40	0.40	0.29	0.47	1.0		
Solar	-0.13	-0.09	-0.13	-0.11	0.21	1.0	
Wind	-0.13	-0.11	-0.17	-0.13	0.23	0.91	1.0
	Perceptions of harm						
	Coal	Natural gas	Oil	Nuclear	Hydro	Solar	Wind
Coal	1.0						
Natural gas	0.48	1.0					
Oil	0.63	0.50	1.0				
Nuclear	0.30	0.32	0.33	1.0			
Hydro	0.30	0.39	0.41	0.28	1.0		
Solar	-0.05	0.25	0.06	0.05	0.30	1.0	
Wind	-0.07	0.17	0.01	-0.03	0.27	0.78	1.0

Source: 2008 MIT/Harvard Energy Survey.

Table A5.1
Determinants of support for future use of energy sources, 2002

	DV: Future use of source (0 to 5: Not use at all to increase a lot)							
	Nuclear	Nuclear	Coal	Natural gas	Oil	Hydro	Solar	Wind
Perceived harm	0.55**	0.35**	0.56***	0.38**	0.36**	0.41**	0.42**	0.45**
(1 to 5: High to none)	(0.063)	(0.061)	(0.050)	(0.051)	(0.054)	(0.047)	(0.054)	(0.073)
Perceived cost	0.14*	0.05	0.08	0.16*	0.13*	0.12*	0.14**	0.07
(1 to 5: Expensive to cheap)	(0.069)	(0.064)	(0.049)	(0.061)	(0.059)	(0.047)	(0.038)	(0.046)
Nuclear safety		0.30**						
		(0.063)						
Nuclear waste storage		0.54**						
		(0.074)						
Global warming concern	-0.12	0.06	-0.01	-0.03	-0.02	0.02	0.16**	0.11
(1 to 4: Not to very)	(0.085)	(0.080)	(0.069)	(0.066)	(0.068)	(0.062)	(0.059)	(0.067)
Electricity bill	-0.05	-0.02	-0.07**	-0.00	-0.04	-0.04	0.02	-0.00
	(0.040)	(0.036)	(0.032)	(0.031)	(0.031)	(0.029)	(0.029)	(0.032)
Age 18–29	-0.47**	-0.39*	0.13	-0.10	-0.01	-0.16	-0.05	-0.20
	(0.169)	(0.154)	(0.136)	(0.132)	(0.133)	(0.125)	(0.123)	(0.139)
Age 60+	-0.05	-0.10	0.19	0.07	0.00	-0.23	0.04	-0.14
	(0.175)	(0.159)	(0.136)	(0.132)	(0.135)	(0.126)	(0.123)	(0.139)
Minority	-0.09	0.04	0.20	0.01	0.38**	-0.05	-0.25*	-0.55**
	(0.153)	(0.139)	(0.123)	(0.119)	(0.123)	(0.115)	(0.112)	(0.126)

(continued)

Table A5.1 (continued)

	DV: Future use of source (0 to 5: Not use at all to increase a lot)							
	Nuclear	Nuclear	Coal	Natural gas	Oil	Hydro	Solar	Wind
Female	-0.00	0.01	-0.12	-0.03	-0.04	-0.04	-0.08	-0.13
	(0.128)	(0.116)	(0.101)	(0.099)	(0.098)	(0.095)	(0.090)	(0.103)
Education	-0.02	-0.03	-0.09*	0.06	-0.11*	-0.06	0.11*	-0.03
	(0.065)	(0.059)	(0.052)	(0.051)	(0.051)	(0.048)	(0.047)	(0.053)
Income	-0.08	-0.08	0.11**	0.05	0.05	-0.01	0.00	0.01
	(0.066)	(0.060)	(0.053)	(0.051)	(0.052)	(0.049)	(0.047)	(0.054)
Democrat	0.22	0.30*	0.21*	-0.01	0.11	0.12	-0.14	-0.02
	(0.156)	(0.142)	(0.124)	(0.120)	(0.121)	(0.114)	(0.111)	(0.128)
Republican	0.35*	0.33*	0.24*	0.07	0.19	0.19	-0.07	-0.09
	(0.161)	(0.146)	(0.128)	(0.124)	(0.125)	(0.118)	(0.115)	(0.130)
Northeast	-0.12	-0.19	0.00	0.10	0.15	-0.11	0.05	-0.10
	(0.187)	(0.170)	(0.151)	(0.146)	(0.146)	(0.138)	(0.133)	(0.152)
South	0.17	0.01	-0.02	0.02	0.25*	-0.03	0.20	-0.16
	(0.162)	(0.147)	(0.130)	(0.126)	(0.127)	(0.119)	(0.116)	(0.132)
West	-0.05	-0.13	-0.42***	0.05	-0.14	0.06	0.40**	0.15
	(0.183)	(0.167)	(0.148)	(0.145)	(0.145)	(0.135)	(0.132)	(0.150)
Intercept	1.23**	-0.71	1.33***	1.30**	1.41**	2.02**	1.11**	2.07**
	(0.445)	(0.449)	(0.397)	(0.381)	(0.389)	(0.379)	(0.369)	(0.471)
Observations	432	429	432	432	430	432	432	432
R-squared	0.327	0.452	0.350	0.226	0.187	0.273	0.306	0.205

Note: Cells contain OLS regression coefficients, with standard errors in parentheses. Significance levels: $^{**}p < 0.01$, $^{*}p < 0.05$.

Table A5.2
Determinants of support for future use of energy sources, 2007

	DV: Future use of source (0 to 5: Not use at all to increase a lot)						
	Coal	Natural gas	Nuclear	Oil	Hydro	Solar	Wind
Perceived harm	0.67**	0.32**	0.64**	0.35**	0.43**	0.40**	0.53**
(1 to 5: High to none)	(0.041)	(0.050)	(0.047)	(0.047)	(0.044)	(0.043)	(0.045)
Perceived cost	0.13**	0.24**	0.20**	0.10*	0.10*	0.09**	0.11**
(1 to 5: Expensive to cheap)	(0.038)	(0.053)	(0.052)	(0.050)	(0.045)	(0.026)	(0.030)
Global warming concern	-0.05	-0.01	-0.02	-0.02	0.12*	0.19**	0.14**
(1 to 4: Not to very)	(0.057)	(0.059)	(0.061)	(0.061)	(0.050)	(0.038)	(0.042)
Electricity bill	0.03	-0.02	0.00	0.00	-0.05*	-0.00	-0.04
	(0.023)	(0.026)	(0.027)	(0.025)	(0.023)	(0.017)	(0.019)
Age 18–29	0.02	-0.07	-0.03	-0.01	-0.03	-0.14	-0.18
	(0.124)	(0.136)	(0.141)	(0.131)	(0.115)	(0.088)	(0.098)
Age 60+	-0.02	-0.08	0.22	-0.22	-0.08	-0.16	-0.15
	(0.108)	(0.120)	(0.133)	(0.117)	(0.108)	(0.082)	(0.092)
Minority	0.09	0.14	-0.16	0.31**	0.23*	-0.02	-0.05
	(0.109)	(0.118)	(0.126)	(0.113)	(0.104)	(0.081)	(0.092)
Female	0.09	0.26**	-0.38**	0.11	-0.03	0.07	0.03
	(0.089)	(0.096)	(0.107)	(0.094)	(0.086)	(0.065)	(0.072)
Education	-0.01	0.01	0.07	-0.06	-0.04	-0.04	0.04
	(0.049)	(0.055)	(0.059)	(0.053)	(0.047)	(0.037)	(0.042)

(continued)

Table A5.2 (continued)

	DV: Future use of source (0 to 5: Not use at all to increase a lot)						
	Coal	Natural gas	Nuclear	Oil	Hydro	Solar	Wind
Income	-0.04	0.12*	0.11	-0.06	0.04	0.05	0.10*
	(0.052)	(0.057)	(0.061)	(0.054)	(0.049)	(0.039)	(0.042)
Democrat	-0.08	0.01	0.26*	0.03	0.18	0.03	-0.02
	(0.104)	(0.113)	(0.120)	(0.109)	(0.100)	(0.076)	(0.085)
Republican	0.20	0.01	0.31*	0.39**	0.09	0.06	0.00
	(0.116)	(0.126)	(0.137)	(0.125)	(0.113)	(0.086)	(0.095)
Northeast	0.03	-0.04	-0.08	0.10	0.13	0.14	0.06
	(0.134)	(0.148)	(0.160)	(0.146)	(0.134)	(0.101)	(0.112)
South	0.11	-0.07	0.03	0.06	0.24*	0.15	0.12
	(0.115)	(0.124)	(0.136)	(0.123)	(0.112)	(0.085)	(0.096)
West	-0.05	0.11	-0.06	-0.07	-0.07	0.09	-0.04
	(0.128)	(0.139)	(0.148)	(0.136)	(0.125)	(0.094)	(0.105)
Intercept	0.22	1.23**	0.55	0.84**	1.36**	1.88**	1.39**
	(0.286)	(0.332)	(0.313)	(0.289)	(0.306)	(0.271)	(0.292)
Observations	444	439	406	454	421	449	436
R-squared	0.525	0.236	0.609	0.249	0.281	0.307	0.379

Note: Cells contain OLS regression coefficients, with standard errors in parentheses. Significance levels: $^{**}p < 0.01$, $^{*}p < 0.05$.

Table A5.3
Determinants of support for future use of energy sources, 2008

	DV: Future use of source (0 to 5: Not use at all to increase a lot)						
	Coal	Natural gas	Nuclear	Oil	Hydro	Solar	Wind
Perceived harm	0.48**	0.34**	0.58**	0.26**	0.47**	0.59**	0.49**
(1 to 5: High to none)	(0.049)	(0.051)	(0.047)	(0.044)	(0.052)	(0.041)	(0.049)
Perceived cost	0.07	0.35**	0.20**	0.28**	0.19**	0.12**	0.11**
(1 to 5: Expensive to cheap)	(0.048)	(0.055)	(0.053)	(0.053)	(0.050)	(0.029)	(0.031)
Global warming concern	-0.17**	0.03	-0.12*	-0.09	0.06	0.08*	0.06
(1 to 4: Not to very)	(0.062)	(0.058)	(0.059)	(0.052)	(0.059)	(0.037)	(0.038)
Electricity bill	-0.01	0.04	0.04	0.02	0.06*	0.02	-0.01
	(0.027)	(0.026)	(0.027)	(0.022)	(0.027)	(0.017)	(0.018)
Age 18–29	-0.06	0.27*	0.05	-0.01	-0.05	-0.01	-0.06
	(0.141)	(0.132)	(0.134)	(0.118)	(0.140)	(0.088)	(0.092)
Age 60+	0.13	0.12	0.24	0.06	0.04	0.12	-0.11
	(0.121)	(0.113)	(0.122)	(0.102)	(0.117)	(0.076)	(0.079)
Minority	0.12	0.03	-0.03	0.09	-0.42**	0.08	-0.03
	(0.124)	(0.119)	(0.124)	(0.105)	(0.124)	(0.080)	(0.084)
Female	0.06	0.07	-0.29**	0.14	-0.11	-0.07	-0.13*
	(0.103)	(0.098)	(0.104)	(0.088)	(0.101)	(0.065)	(0.067)
Education	0.13*	0.02	0.13*	-0.06	0.13*	0.16**	0.07
	(0.058)	(0.054)	(0.057)	(0.049)	(0.057)	(0.037)	(0.039)

(continued)

Table A5.3 (continued)

	DV: Future use of source (0 to 5: Not use at all to increase a lot)						
	Coal	Natural gas	Nuclear	Oil	Hydro	Solar	Wind
Income	-0.08	0.08	0.09	-0.14**	-0.15*	-0.01	0.01
	(0.062)	(0.058)	(0.061)	(0.052)	(0.063)	(0.039)	(0.040)
Democrat	-0.19	0.19	-0.03	-0.10	0.02	0.07	-0.07
	(0.237)	(0.220)	(0.227)	(0.229)	(0.230)	(0.157)	(0.156)
Republican	-0.05	0.25	-0.08	0.08	0.07	0.02	-0.11
	(0.237)	(0.220)	(0.228)	(0.230)	(0.232)	(0.156)	(0.157)
Northeast	-0.32*	-0.03	0.11	-0.16	0.29	-0.11	-0.12
	(0.156)	(0.148)	(0.155)	(0.133)	(0.154)	(0.100)	(0.103)
South	-0.23	0.27*	0.06	0.04	0.04	0.02	-0.04
	(0.138)	(0.130)	(0.139)	(0.119)	(0.139)	(0.088)	(0.091)
West	-0.12	0.36*	0.03	0.21	0.06	0.13	0.09
	(0.152)	(0.147)	(0.150)	(0.130)	(0.147)	(0.095)	(0.097)
Intercept	1.20**	-0.05	0.81*	0.92*	0.28	0.52	1.75**
	(0.421)	(0.403)	(0.381)	(0.375)	(0.439)	(0.300)	(0.321)
Observations	436	433	421	451	422	461	451
R-squared	0.313	0.290	0.577	0.240	0.331	0.492	0.347

Note: Cells contain OLS regression coefficients, with standard errors in parentheses. Significance levels: $^{**}p < 0.01$, $^{*}p < 0.05$.

Table A5.4
Determinants of support for future use of energy sources, 2011

	DV: Future use of source (0 to 5: Not use at all to increase a lot)						
	Coal	Natural gas	Nuclear	Oil	Hydro	Solar	Wind
Perceived harm	0.55**	0.40**	0.64**	0.42**	0.31**	0.52**	0.46**
(1 to 5: High to none)	(0.038)	(0.038)	(0.033)	(0.035)	(0.040)	(0.060)	(0.057)
Perceived cost	0.21**	0.38**	0.21**	0.30**	0.18**	0.23**	0.32**
(1 to 5: Expensive to cheap)	(0.038)	(0.039)	(0.036)	(0.036)	(0.039)	(0.031)	(0.031)
Global warming concern	-0.17**	-0.09*	-0.01	-0.13**	-0.08*	0.19**	0.27**
(1 to 4: Not to very)	(0.046)	(0.038)	(0.041)	(0.039)	(0.040)	(0.041)	(0.043)
Age 18–29	-0.02	-0.25*	-0.14	0.09	-0.26*	-0.16	-0.05
	(0.109)	(0.101)	(0.104)	(0.095)	(0.102)	(0.109)	(0.109)
Age 60+	0.26**	0.07	0.04	0.08	-0.05	0.02	-0.09
	(0.099)	(0.089)	(0.096)	(0.087)	(0.096)	(0.099)	(0.099)
Minority	0.49**	0.24*	0.11	0.24**	0.20*	0.16	0.08
	(0.107)	(0.095)	(0.099)	(0.092)	(0.099)	(0.105)	(0.104)
Female	0.03	-0.22**	-0.20*	0.10	-0.14	-0.20*	-0.17*
	(0.086)	(0.077)	(0.084)	(0.075)	(0.082)	(0.084)	(0.083)
Education	-0.00	0.04	0.06	0.03	-0.00	0.11**	0.08*
	(0.033)	(0.029)	(0.032)	(0.028)	(0.032)	(0.032)	(0.033)

(continued)

Table A5.4 (continued)

	DV: Future use of source (0 to 5: Not use at all to increase a lot)						
	Coal	Natural gas	Nuclear	Oil	Hydro	Solar	Wind
Income	0.04	-0.03	0.02	0.03	-0.02	-0.00	0.03
	(0.044)	(0.040)	(0.041)	(0.039)	(0.042)	(0.044)	(0.044)
Democrat	0.01	0.39**	0.06	0.16	0.00	0.03	-0.05
	(0.145)	(0.127)	(0.132)	(0.126)	(0.139)	(0.141)	(0.144)
Republican	0.16	0.49**	0.41**	0.36**	-0.18	-0.39**	-0.16
	(0.151)	(0.131)	(0.139)	(0.131)	(0.144)	(0.146)	(0.150)
Northeast	0.12	-0.09	0.14	-0.01	-0.11	-0.27*	-0.38**
	(0.130)	(0.118)	(0.123)	(0.114)	(0.124)	(0.128)	(0.129)
South	0.04	-0.12	0.17	-0.02	-0.16	-0.20	-0.40**
	(0.108)	(0.096)	(0.102)	(0.094)	(0.103)	(0.106)	(0.106)
West	-0.00	-0.18	0.09	-0.03	-0.26*	-0.11	-0.20
	(0.123)	(0.110)	(0.119)	(0.108)	(0.119)	(0.122)	(0.124)
Intercept	0.74*	0.72**	0.10	0.46	2.19**	0.35	0.22
	(0.308)	(0.277)	(0.246)	(0.255)	(0.300)	(0.360)	(0.322)
Observations	726	733	706	737	702	764	754
R-squared	0.437	0.431	0.608	0.441	0.176	0.291	0.334

Note: Cells contain OLS regression coefficients, with standard errors in parentheses. Significance levels: $^{**}p < 0.01$, $^{*}p < 0.05$.

Table A7.1
Public concern about climate change

There is a lot of talk about global warming caused by carbon dioxide emissions from burning coal, natural gas, and oil to generate electricity. In thinking about the important problems facing the United States and about your personal situation. Is this problem: global warming	2002†
Very important	43%
Somewhat important	42%
Not very important	11%
Not important at all	4%

From what you know about global warming, which of the following statements comes closest to your opinion?	2003†	2006†
Global warming has been established as a serious problem and immediate action is necessary.	17%	28%
There is enough evidence that global warming is taking place and some action should be taken.	36%	34%
We don't know enough about global warming and more research is necessary before we take any actions.	24%	18%
Concern about global warming is unwarranted.	7%	6%
No opinion	16%	14%

From what you know about global climate change or global warming, which one of the following statements comes closest to your opinion?	2006‡	2007†	2008†	2009‡
Immediate and drastic action is necessary	35%	19%	25%	23%
Some action is necessary	27%	43%	43%	33%
More research is needed before action is taken	22%	28%	21%	20%
This is not a serious problem	16%	10%	11%	25%

(continued)

Table A7.1 (continued)

From what you know about global climate change, which of the following statements comes closest to your opinion?	2010‡	2011‡
Global climate change has been established as a serious problem, and immediate action is necessary.	27%	27%
There is enough evidence that climate change is taking place, and some action should be taken.	29%	31%
We don't know enough about global climate change, and more research is necessary before taking any actions.	21%	21%
Concern about global climate change is exaggerated. No action is necessary.	17%	14%
Global climate change is not occurring; this is not a real issue.	6%	6%

Sources: †MIT/Harvard Energy Surveys; ‡Cooperative Congressional Election Study.
Note: MIT 2008 wording: Which of the following do you think best describes your view about global warming?

Table A7.2
Willingness to pay to resolve climate change

	2003	2006	2007	2008	2011	2013
Amount per month	Percent responding yes					
$0	24	21	26	23	39	31
$5	22	18	28	29	22	15
$10	32	29	20	19	13	10
$15	-	-	11	11	10	30
$25	13	13	9	9	8	9
$50	4	8	2	3	3	2
$75	-	-	1	1	1	1
$100	5	10	4	5	5	2
Estimated average WTP	$14.44	$21.35	$12.85	$14.49	$11.48	$12.11
Estimated median WTP	$10.00	$10.00	$5.00	$5.00	$5.00	$10.00

Source: MIT/Harvard Energy Surveys.

Table A8.1
Determinants of support for climate policy options, 2011

	DV: Support for policy (1 = Support and 0 = Oppose)				
	Regulatory cap	Cap and trade	Carbon tax	Carbon tax + Income tax cut	Carbon tax + Deficit reduction
Global warming concern	0.12**	0.12**	0.09**	0.05**	0.04**
(1 to 5: Not to very)	(0.009)	(0.017)	(0.008)	(0.011)	(0.012)
Environmental harm traditional fuels	-0.11**	-0.05	-0.08**	-0.02	-0.04*
(Higher values mean less harm)	(0.014)	(0.026)	(0.013)	(0.017)	(0.018)
Economic cost traditional fuels	-0.05**	0.00	0.01	-0.06**	-0.03
(Higher values mean cheaper)	(0.011)	(0.023)	(0.010)	(0.014)	(0.015)
Environmental harm alternative energy	0.06**	0.01	-0.02*	0.06**	0.02
(Higher values mean less harm)	(0.010)	(0.020)	(0.010)	(0.013)	(0.013)
Economic cost alternative fuels	0.03**	0.04	-0.00	0.06**	0.04**
(Higher values mean cheaper)	(0.011)	(0.020)	(0.010)	(0.013)	(0.014)
Electricity bill	-0.01*	-0.01	-0.00	0.00	0.00
	(0.005)	(0.008)	(0.004)	(0.006)	(0.006)
Miles driven (log)	0.00	-0.00	-0.01**	-0.00	-0.00
	(0.003)	(0.006)	(0.003)	(0.004)	(0.004)
Age 18–29	-0.01	-0.09*	0.03	0.11**	0.09**
	(0.023)	(0.043)	(0.021)	(0.028)	(0.029)
Age 60+	0.08**	-0.13**	-0.01	-0.14**	-0.02
	(0.021)	(0.039)	(0.019)	(0.026)	(0.027)
Minority	-0.02	0.13**	0.02	0.05	0.07**
	(0.021)	(0.038)	(0.019)	(0.026)	(0.027)
Female	-0.00	0.09*	-0.05**	-0.01	-0.01
	(0.018)	(0.034)	(0.017)	(0.022)	(0.023)
Education	0.01*	-0.02	0.02**	-0.02**	0.00
	(0.007)	(0.013)	(0.006)	(0.008)	(0.009)
Income	-0.09	0.20	0.28**	-0.15	-0.13
	(0.076)	(0.136)	(0.070)	(0.095)	(0.098)
Political ideology	-0.06**	-0.06*	-0.08**	-0.02	-0.03
	(0.012)	(0.024)	(0.011)	(0.015)	(0.016)

Table A8.1 (continued)

	DV: Support for policy (1 = Support and 0 = Oppose)				
	Regulatory cap	Cap and trade	Carbon tax	Carbon tax + Income tax cut	Carbon tax + Deficit reduction
Democrat	-0.05	0.09	-0.02	0.05	0.07
	(0.029)	(0.054)	(0.026)	(0.035)	(0.037)
Republican	-0.11**	-0.10	-0.09**	0.04	-0.04
	(0.031)	(0.059)	(0.028)	(0.038)	(0.039)
Northeast	0.02	0.09	0.05*	0.04	0.06
	(0.027)	(0.050)	(0.025)	(0.033)	(0.034)
South	-0.02	-0.03	0.05*	0.02	0.01
	(0.023)	(0.040)	(0.021)	(0.028)	(0.029)
West	-0.02	0.01	0.04	-0.04	0.02
	(0.026)	(0.049)	(0.024)	(0.033)	(0.034)
Intercept	0.43**	0.48**	0.45**	0.43**	0.32**
	(0.086)	(0.164)	(0.079)	(0.106)	(0.110)
Observations	1,936	659	1,935	1,915	1,915
R-squared	0.410	0.394	0.343	0.151	0.113

Source: MIT/Harvard Energy Surveys.
Note: Cells contain OLS regression coefficients, with standard errors in parentheses. Models also include imputation for missing values on measures of harms and costs, and dummy variables for missing data on income, political ideology, electricity bill, and miles drive. Significance levels: $^{**}p < 0.01$, $^{*}p < 0.05$.

Notes

1 The Energy Challenge

1. These are California plus the nine states of the Regional Greenhouse Gas Initiative (Connecticut, Delaware, Maine, Maryland, Massachusetts, New Hampshire, New York, Rhode Island, and Vermont).

2. U.S. Energy Information Administration, "Projected Retirements of Coal-Fired Power Plants," July 31, 2012, http://www.eia.gov/todayinenergy/detail.cfm?id=7330 (accessed August 26, 2013).

3. U.S. Energy Information Administration, *Annual Energy Outlook 2012*, DOE/EIA-0383(2012), June 2012, U.S. Department of Energy, Washington, DC, 86.

4. The largest coal power plants have capacity to produce approximately 2 gigawatts of electricity.

5. The Three Mile Island nuclear power plant is rated as 800 megawatts or 0.8 gigawatts.

6. U.S. Energy Information Administration, *Annual Energy Outlook 2012*, Table 4.1.

7. Congress created the Federal Power Commission to deal with the problems of interstate transmission of electricity and the regulation of hydroelectricity in the 1920s. Congress created the Department of Energy and reorganized the FPC into the Federal Energy Regulatory Commission (FERC) during the 1970s. For additional discussion, see Michael J. Graetz, *The End of Energy: The Unmaking of America's Environment, Security, and Independence* (Cambridge, MA: MIT Press, 2011).

8. Paul Joskow, "Energy Policies and Their Consequences 25 Years Later," *Energy Journal* 24, no. 4 (2003): 17–49.

9. Don Fullerton and Catherine Wolfram, eds., *The Design and Implementation of U.S. Climate Policy* (Chicago: University of Chicago Press, 2012).

10. Several recent books have eloquently discussed the policy challenges of moving toward a lower-carbon energy economy in light of climate change. See, for example, John M. Deutch, *The Crisis in Energy Policy* (Cambridge, MA: Harvard University Press, 2001); Kelly Sims Gallagher, *Acting in Time on Energy*

Policy (Washington, DC: Brookings Institution Press, 2009); Michael J. Graetz, *The End of Energy: The Unmaking of America's Environment, Security, and Independence* (Cambridge, MA: MIT Press, 2011); Richard K. Lester and David M. Hart, *Unlocking Energy Innovation: How America Can Build a Low-Cost, Low-Carbon Energy System* (Cambridge, MA: MIT Press, 2012); and Charles Weiss and William B. Bonvillian, *Structuring and Energy Revolution* (Cambridge, MA: MIT Press, 2009).

11. R. L. Lehr, W. Guild, D. L. Thomas, and B. G. Swezey, "Listening to Customers: How Deliberative Polling Helped Build 1,000 MB of New Renewable Energy Projects in Texas," Technical Report, NREL/TP-620-33177, National Renewable Electricity Laboratory, Golden, CO, June 2003.

12. William A. Freudenburg and Eugene A. Rosa, *Public Reaction to Nuclear Power: Are There Critical Masses?* (Boulder, CO: Westview Press, 1984); Deborah Lynn Guber, *The Grassroots of a Green Revolution: Polling America on the Environment* (Cambridge, MA: MIT Press, 2003); Paul Slovic, "Perception of Risk," *Science* 236 (1987): 280–285; and Yutaka Tanaka, "Major Psychological Factors Determining Public Acceptance of the Siting of Nuclear Facilities." *Journal of Applied Social Psychology* 34 (2004): 1147–1165.

13. Slovic, "Perception of Risk"; Paul Slovic, "Perception of Risk and the Future of Nuclear Power," *Physics and Society* 23, no. 1 (1994): 3–5; and Paul Slovic, James H. Flynn, and Mark Layman, "Perceived Risk, Trust, and the Politics of Nuclear Waste," *Science* 254, no. 5038 (1991): 1603–1607.

14. Search was conducted on August 16, 2013.

15. For a good summary, see Matthew C. Nisbet and Teresa Myers, "The Polls—Trends Twenty Years of Public Opinion about Global Warming," *Public Opinion Quarterly* 71, no. 3 (2007): 444–470.

16. See, for example, Thomas Dietz, Amy Dan, and Rachael Shwom, "Support for Climate Change Policy: Social Psychological and Social Structural Influences," *Rural Sociology* 72, no. 2 (2007): 185–214; Anthony Leiserowitz, "Climate Change Risk Perception and Policy Preferences: The Role of Affect, Imagery, and Values," *Climatic Change* 77, nos. 1–2 (2006): 45–72; Andreas Nilsson, Chris von Borgstede, and Anders Biel, "Willingness to Accept Climate Change Strategies: The Effect of Values and Norms," *Journal of Environmental Psychology* 24, no. 3 (2004): 267–277; Robert E. O'Connor, Richard J. Bord, and Ann Fisher, "Risk Perceptions, General Environmental Beliefs, and Willingness to Address Climate Change," *Risk Analysis* 19, no. 3 (1999): 461–471; and Rachael Shwom, David Bidwell, Amy Dan, and Thomas Dietz, "Understanding U.S. Public Support for Domestic Climate Change Policies," *Global Environmental Change* 20, no. 3 (2010): 472–482.

17. Jon A. Krosnick, "The Climate Majority," *The New York Times*, June 8, 2010; Jon A. Krosnick and Bo MacInnis, "Does the American Public Support Legislation to Reduce Greenhouse Gas Emissions," *Daedalus* 142, no. 1 (2013): 26–39.

18. A. Leiserowitz, E. Maibach, C. Roser-Renouf, and J. Hmielowski, "Global Warming's Six Americas 2009: An Audience Segmentation Analysis, May 2009," Yale University and George Mason University, Yale Project on Climate Change Communication, New Haven, CT, 2012; and A. Leiserowitz, E. Maibach, C. Roser-Renouf, and J. Hmielowski, "Global Warming's Six Americas, March 2012 & Nov. 2011," Yale University and George Mason University, Yale Project on Climate Change Communication, New Haven, CT, 2012.

2 Energy Choices

1. It was not always that way. For decades most states and locales provided power through public utilities—local, government-created monopolies. In the 1980s, many of these were dismantled and U.S. electricity markets became privatized in many states. See Paul L. Joskow and Richard Schmalensee, *Markets for Power* (Cambridge, MA: MIT Press, 1988).

2. This is not to suggest that there are no near-term impacts from climate change, just that they are conceptually separable from the longer impacts.

3. Carbon capture and storage remains a nascent technology, and there remain significant hurdles to employing it at a large scale.

4. Graetz, *The End of Energy*.

5. U.S Energy Information Administration, *Annual Energy Review 2009*, August 2010, http://www.eia.doe.gov/aer/overview.html (accessed August 26, 2013).

6. International Monetary Fund, World Economic Outlook Database, October 2010, http://www.imf.org/external/pubs/ft/weo/2010/02/weodata/index.aspx (accessed August 26, 2013).

7. This discussion is adapted largely from Duke Energy, "How Do Coal-Fired Plants Work?," http://www.duke-energy.com/about-energy/generating-electricity/coal-fired-how.asp (accessed August 26, 2013).

8. Overnight capital costs refers to the cost of building a power plant assuming that the entire process from planning through completion could be completed in a single day. This concept is useful because it avoids having to make assumptions about estimated costs from delays and financing.

9. U.S. Energy Information Administration, "Levelized Cost of New Generation Resources," in the *Annual Energy Outlook 2012*, July 2012.

10. The estimates generated by the EIA do not reflect federal or state subsidies or tax incentives in place to encourage certain technologies.

11. This lower cost of natural gas compared to coal and nuclear power is reflected in recent trends in electricity generation. From 2008 to 2009, generation from natural gas increased by 4.3 percent, contributing to its highest share of the national total in four decades. Meanwhile, generation from coal dropped by 11.6 percent over this same period (U.S. Energy Information Administration, *Electric Power Annual 2009*, DOE/EIA-0384(2009), U.S. Department of Energy, Washington, DC, August 2010, 1).

12. See, for example, Paul N. Leiby, "Estimating the Energy Security Benefits of Reduced U.S. Oil Imports," Final Report, ORNL/TM-2007/028, Oak Ridge National Laboratory, revised March 14, 2008.

13. U.S. Environmental Protection Agency, "EPA, Mercury and Air Toxics Standards," http://www.epa.gov/airquality/powerplanttoxics/ (accessed April 1, 2013).

14. National Research Council, *Hidden Costs of Energy* (Washington, DC: National Academies Press, 2010).

15. Michael Greenstone and Adam Looney, "A Strategy for America's Energy Future: Illuminating Energy's Full Costs," The Hamilton Project, Washington, DC, 2011.

16. Specifically, they use estimates from Yangbo Du and John E. Parsons, "Update on the Costs of Nuclear Power," Working Paper 09-994, MIT Center for Energy and Environmental Policy Research, May 2009.

17. Estimates of the social costs of carbon range considerably, which is not surprising given all of the uncertainties about the consequences of climate change as well as different modeling assumptions, such as discount rates. The $21 per ton used by Greenstone and Looney reflects the central estimate from the federal government's 2010 Interagency Working Group on the Social Costs of Carbon. In May 2013, the federal government increased the estimate it uses in rulemaking from $21 to $37 a ton for 2015. A 2011 study by Frank Ackerman and Elizabeth A. Stanton for the Economics for Equity & Environment estimated the social costs of carbon to be about $900/ton, about forty-five times the estimate of used by Greenstone and Looney. See http://www.e3network.org/papers/Climate_Risks_and_Carbon_Prices_executive-summary_full-report_comments.pdf. For further discussion see, among others, Richard S. J. Tol, "The Marginal Damage Costs of Carbon Dioxide Emissions: An Assessment of the Uncertainties," *Energy Policy* 33, no. 16 (2005): 2064–2074; David Pearce, "The Social Cost of Carbon and Its Policy Implications," *Oxford Review of Economic Policy* 19, no. 3 (2003): 362–384.

18. Greenstone and Looney separately estimate the costs of conventional combustion natural gas turbines at 12.2 cents per kWh.

19. Nuclear Energy Institute, "Nuclear Waste: Amounts & On-Site Storage," http://www.nei.org/Knowledge-Center/Nuclear-Statistics/On-Site-Storage-of-Nuclear-Waste (accessed December 16, 2013).

20. MIT, "The Future of Nuclear Power," Cambridge, MA, 2003, 10.

21. Nicholas Stern, *The Economics of Climate Change: The Stern Review* (Cambridge: Cambridge University Press, 2006).

22. U.S. Environmental Protection Agency, "National Greenhouse Gas Emissions Data," 2011, http://www.epa.gov/climatechange/emissions/usinventoryreport.html (accessed December 20, 2013).

23. Joseph E. Aldy, "Promoting Clean Energy in the American Power Sector," The Hamilton Project, Washington, DC, May 2011.

24. Specifically, we use the Greenstone and Looney ("A Strategy for America's Energy Future," Table 1) estimates, which characterize the noncarbon social costs of nuclear wind and solar power as "unable to characterize" or zero.

3 What People Want

1. Lehr et al., "Listening to Customers"; and J. Fishkin, "Deliberative Polling: Toward a Better-Informed Democracy," Center for Deliberative Democracy, 2010, cdd.stanford.edu/polls/docs/summary; Kate Galbraith and Asher Price, *The Great Texas Wind Rush: How George Bush, Ann Richards, and a Bunch of Tinkerers Helped the Oil and Gas State Win the Race to Wind Power* (Austin: University of Texas Press, 2013).

2. "The Frontier Spirit," *Economist*, May 14, 1998, http://www.economist.com/node/128792 (accessed December 20, 2013).

3. Clifford Krauss, "Move Over, Oil, There's Money in Texas Wind," *New York Times*, February 23, 2008. http://www.nytimes.com/2008/02/23/business/23wind.html?_r=2&hp&oref=slogin& (accessed December 20, 2013).

4. Ibid.

5. For an overview of what energy survey research has focused on, see Toby Bolsen and Fay Lomax Cook, "The Polls—Trends: Public Opinion on Energy Policy: 1974–2006," *Public Opinion Quarterly* 72 (2008): 364–388; and Eugene Rosa and Riley Dunlap, "Nuclear Power: Three Decades of Public Opinion," *Public Opinion Quarterly* 58 (1994): 295–324.

6. In recent years, there has been growing attention to public attitudes toward wind, but most of the studies are not national in scale and focus on public opinion in European nations. See, for example, David Bidwell, "The Role of Values in Public Beliefs and Attitudes towards Commercial Wind Energy," *Energy Policy* 58 (2013): 189–199; Jeremy Firestone and Willett Kempton, "Public Opinion about Large Offshore Wind Power: Underlying Factors," *Energy Policy* 35, no. 3 (2007): 1584–1598; Søren Krohn and Steffen Damborg, "On Public Attitudes towards Wind Power," *Renewable Energy* 16, no. 1 (1999): 954–960; Maarten Wolsink, "Wind Power and the NIMBY-Myth: Institutional Capacity and the Limited Significance of Public Support," *Renewable Energy* 21, no. 1 (2000): 49–64; and Maarten Wolsink, "Wind Power Implementation: The Nature of Public Attitudes: Equity and Fairness Instead of 'Backyard Motives,'" *Renewable and Sustainable Energy Reviews* 11 (2007): 118–1207.

7. Barbara C. Farhar, Charles T. Unseld, Rebecca Vories, and Robin Crews, "Public Opinion about Energy," *Annual Review of Energy* 5 (1980): 141–172.

8. Some of the best data come from a series of surveys conducted by Cambridge Research that asked Americans to indicate their support for several options for dealing with the energy crisis, including the following: "increasing strip-mining for coal, even if it means damaging the environment," and "expanding offshore drilling for oil and natural gas," "nationalization of all the big oil companies," "creating a federal oil reserve where the United States would store up large

amounts of oil as protection against a cutoff of foreign oil," "converting electric utilities to burn coal, even if coal burning means dirtier air," "a national energy corporation that would conduct research and also own energy resources on public land," "a tariff tax on imported oil that would raise the price of gasoline 10 to 15 cents per gallon, if it would end our dependence on foreign oil," "setting up a new federal board with the power to overrule environmental regulations and other laws that slow down the building of important energy projects," and "ending price controls on natural gas."

9. Even before the Three Mile Island accident, nuclear power was beginning to be increasingly perceived in a negative light. For a comprehensive discussion, see Frank R. Baumgartner and Bryan D. Jones, *Agendas and Instability in American Politics* (Chicago: University of Chicago Press, 1993).

10. U.S. Nuclear Regulatory Commission, "Three Mile Island Accident," 2013, http://www.nrc.gov/reading-rm/doc-collections/fact-sheets/3mile-isle.pdf (accessed December 16, 2013).

11. Bolsen and Cook, "The Polls—Trends."

12. Barbara C. Farhar, "Trends in U.S. Public Perceptions and Preferences on Energy and Environmental Policy," *Annual Review of Energy and the Environment* 19 (1994): 211–239; and Farhar et al., "Public Opinion about Energy."

13. The response data for the 1991 Yankelovich survey were as follows: coal—rely more (43%), rely less (49%), not sure (8%); oil—rely more (27%), rely less (68%), not sure (5%); and nuclear power—rely more (44%), rely less (48%), not sure (8%).

14. U.S. Energy Information Administration, *Electric Power Annual 2011*, U.S. Department of Energy, Washington, DC, January 2013.

15. Stephen Ansolabehere and Brian Schaffner, "Does Survey Mode Still Matter? Findings from a 2010 Multi-Mode Comparison," forthcoming in *Political Analysis*.

16. Knowledge Networks was acquired by GfK Custom Research in December 2011.

17. Douglas Rivers, "Sampling for Web Surveys," paper presented at the 2007 Joint Statistical Meetings, Salt Lake City, UT, 2007; and Lynn Vavreck and Douglas Rivers, "The 2006 Cooperative Congressional Election Study," *Journal of Elections, Public Opinion and Parties* 18 (2008): 355–366.

18. As we discuss in chapter 6, this question was part of a randomized experiment in each of the four surveys, in which subsamples of the survey received different information prior to being asked this question. The responses described here reflect just those from the control groups in the experiments (i.e., the part of the samples that did not receive any information), although the general picture looks the same for the full sample.

19. John Broder, "Nebraska Governor Approves Keystone XL Route," *New York Times*, January 22, 2013.

20. Lucas Davis, "The Effect of Power Plants on Local Housing Prices and Rents: Evidence from Restricted Census Microdata," *Review of Economics and Statistics* 93, no. 4 (2011): 1391–1402.

21. See, for example, Susan Hunter and Kevin M. Leyden, "Beyond NIMBY: Explaining Opposition to Hazardous Waste Facilities," *Policy Studies Journal* 23 (1995): 601–619; Hank Jenkins-Smith and Howard Kunreuther, "Mitigation and Benefits Measures as Policy Tools for Siting Potentially Hazardous Facilities: Determinants of Effectiveness and Appropriateness," *Risk Analysis* 21 (2001): 371–382; Philip H. Pollock III, M. Elliot Vittes, and Stuart A. Lilie, "Who Says It's Risky Business? Public Attitudes toward Hazardous Waste Facility Siting," *Polity* 24 (1992): 499–513; and Kent E. Portney, *Siting Hazardous Waste Treatment Facilities: The NIMBY Syndrome* (New York: Auburn House, 1991).

22. Tanaka, "Major Psychological Factors Determining Public Acceptance of the Siting of Nuclear Facilities"; Krohn and Damborg, "On Public Attitudes towards Wind Power"; and Charles R. Warren, Carolyn Lumsden, Simone O'Dowd, and Richard V. Birnie, "'Green on Green': Public Perceptions of Wind Power in Scotland and Ireland," *Journal of Environmental Planning and Management* 48 (2005): 853–875.

23. Rosa and Dunlap, "Poll Trends."

24. Freudenburg and Rosa, *Public Reaction to Nuclear Power*; and Christian Joppke, *Mobilizing against Nuclear Energy: A Comparison of Germany and the United States* (Berkeley: University of California Press, 1993).

25. These findings are particularly strong in the 2002, 2007, and 2008 surveys, but soften somewhat in the 2011 sample.

26. Warren et al., "'Green on Green.'"

27. Stephen Ansolabehere and David M. Konisky, "Public Attitudes toward Construction of New Power Plants," *Public Opinion Quarterly* 73, no. 3 (2009): 566–577.

28. Bolsen and Cook, "The Polls—Trends."

29. Deborah Haber, "Seventies Oil Crisis Was a 'Perfect Storm' for U.S.," MIT News, March 23, 2007, http://newsoffice.mit.edu/2007/seventies-oil-crisis-was-perfect-storm-us (accessed April 30, 2014).

4 Price and Consequence

1. Eric R. A. N. Smith, *Energy, the Environment, and Public Opinion* (Lanham, MD: Rowman and Littlefield Publishers, 2003). See also Benjamin Page and Robert Shapiro, *The Rational Public: Fifty Years of Trends in Americans' Policy Preferences* (Chicago: University of Chicago Press, 1992), especially pp. 150–156.

2. Bolsen and Cook, "The Polls—Trends"; Farhar, "Trends in U.S. Public Perceptions and Preferences on Energy and Environmental Policy"; Farhar et al., "Public Opinion about Energy."

3. Gallup, “Americans Still Divided on Energy-Environment Trade-Off,” April 10, 2013, http://www.gallup.com/poll/161729/americans-divided-energy-environment-trade-off.aspx (accessed September 18, 2013).

4. Robert Tuttle and Ola Galal, “Oil Ministers See Demand, Prices Rising, Undeterred by Greek Debt Crisis,” *Bloomberg News*, May 10, 2010.

5. A similar shift occurred in Gallup’s question that asks people to make a trade-off between protecting the environment and economic growth. Historically, a majority of the public has prioritized environmental protection, but this changed in 2008, and since then a majority of the public has prioritized economic growth. See Gallup, “More Americans Still Prioritize Economy over Environment,” April 3, 2013, http://www.gallup.com/poll/161594/americans-prioritize-economy-environment.aspx (accessed August 27, 2013)

6. Howard Schuman and Stanley Presser, “The Open and Closed Question,” *American Sociological Review* 44 (October 1979): 692–712.

7. Stephen Ansolabehere, Marc Meredith, and Erik Snowberg, “Asking about Numbers: Why and How,” *Political Analysis* 21, no. 1 (2013): 48–69.

8. Data reported in Farhar, “Trends in U.S. Public Perceptions and Preferences on Energy and Environmental Policy.”

9. The environmental harm most typically associated with wind power is bird kills, while with solar power it is land use given the large footprint of utility-scale solar installation.

10. See, for example, Ernest Moniz, “Why We Still Need Nuclear Power: Making Clean Energy Safe and Affordable,” *Foreign Affairs*, November/December 2011.

5 Why Do People Hate Coal and Love Solar?

1. For the most comprehensive account of energy preferences to date, see Smith, *Energy, the Environment, and Public Opinion*. For an outstanding treatment of the environmental side of the question, see Judy Layzer, *The Environmental Case*, 3rd ed. (Washington DC: CQ Press, 2011). On moral and social values, see Robert Worcester, “Public Opinion and the Environment,” *The Political Quarterly* 68 (1997): 160–173.

2. For a discussion of the relevance of culture and views of technology generally, see Mary Douglas and Aaron Wildavsky, *Risk and Culture* (Berkeley: University of California Press, 1992).

3. See Sheldon Kamieniecki, “Political Parties and Environmental Policy,” in *Environmental Politics and Policy*, 2nd ed., ed. James P. Lester, 146–167 (Durham, NC: Duke University Press, 1995); and Guber, *The Grassroots of a Green Revolution*.

4. U.S. Environmental Protection Agency, “EPA Analysis of the American Power Act in the 111th Congress,” June 2010, http://www.epa.gov/climatechange/Downloads/EPAactivities/EPA_APA_Analysis_6-14-10.pdf (accessed August 17, 2013).

5. Congressional Budget Office, "The Estimated Costs to Households from the Cap-and-Trade Provisions of H.R. 2454," June 19, 2009, http://www.cbo.gov/sites/default/files/cbofiles/ftpdocs/103xx/doc10327/06-19-capandtradecosts.pdf (accessed August 17, 2013).

6. William W. Beach, David Kreutzer, Karen Campbell, and Ben Lieberman, "Son of Waxman-Markey: More Politics Makes for a More Costly Bill," Heritage Foundation WebMemo No. 2450, May 18, 2009.

7. See, for example, Frank Ackerman and Lisa Heinzerling, *Priceless: On Knowing the Price of Everything and the Value of Nothing* (New York: The New Press, 2004).

8. We also included a separate measure of political ideology in unreported models, but do not include it in the results we discuss, because it was not available in each survey. Including or omitting political ideology does not change our substantive findings in any meaningful way.

9. See, for example, Christopher Jan Carman, "Dimensions of Environmental Policy Support in the United States," *Social Science Quarterly* 79, no. 4 (1998): 717–733; Riley E. Dunlap, Chenyang Xiao, and Aaron M. McCright, "Politics and Environment in America: Partisan and Ideological Cleavages in Public Support for Environmentalism," *Environmental Politics* 10, no. 4 (2001): 23–48; Guber, *The Grassroots of a Green Revolution*; Stephen L. Klineberg, Matthew McKeever, and Bert Rothenbach, "Demographic Predictors of Environmental Concern: It Does Make a Difference How It's Measured," *Social Science Quarterly* 79, no. 4 (1998): 734–753; and David M. Konisky, Jeffrey Milyo, and Lilliard E. Richardson Jr., "Environmental Policy Attitudes: Issues, Geographical Scale, and Political Trust," *Social Science Quarterly* 89, no. 5 (2008): 1066–1085.

10. It is important to note that the survey question used to measure the dependent variable in the 2002 survey was part of a framing experiment in which groups of respondents received different information immediately prior to being asked this question. We discuss this experiment in detail in the next chapter, and the results presented in table 5.1 are only for the control group in the experiment—that is, those individuals that did not receive any information prior to responding to this question. For this reason, the sample size is small (about 435 people) in the models reported. We did estimate the same models for the entire sample, including dummy variables for the each of the treatment groups in the experiment, and the main results are substantively identical.

11. We also included political ideology and risk attitudes in original analysis, but we do not include them here because they are not available in all surveys.

12. To simplify interpretation, we report estimates from OLS regression models, which treat the ordered response categories as if they were continuous. We also estimated ordered logit models as a robustness check, and the results are substantively the same.

13. When we added separate measures of perceptions of the safety of nuclear power plants (more precisely, perceptions of the likelihood of a major accident)

and of the capacity to safely storage nuclear waste, the effect of perceived harms remains important, although the size of the coefficient is a little smaller, while the coefficient on perceived costs was no longer statistically significant. As would be expected, those individuals who least feared an accident and believed that waste could be safely stored were more likely to support the increased use of nuclear power.

14. Standardized regression coefficients allow for more direct comparison of the measures used. One can interpret the expected change in the dependent variable for each standard deviation increase in the measure of interest.

15. Ansolabehere, Meredith, and Snowberg, "Asking about Numbers."

16. Thomas Dietz, Paul C. Stern, and Elke U. Weber, "Reducing Carbon-Based Energy Consumption through Changes in Household Behavior" *Daedalus* 142 (Winter 2013): 78–89. There are many excellent treatments of public opinion and information, but we highly recommend John Zaller, *The Nature and Origin of Public Opinion* (Cambridge: Cambridge University Press, 1991); Ted Carmines and James Stimson, *Issue Evolution* (Princeton: Princeton University Press, 1989; and Page and Shapiro, *The Rational Public.*

17. See Shanto Iyengar and Donald R. Kinder, *News that Matters* (Chicago: University of Chicago Press, 1989).

18. Paul Slovic, "Perception of Risk and the Future of Nuclear Power," *Arizona Journal of International and Comparative Law* 9 (1992): 191–198; and Elke U. Weber, "Experience-Based and Description-Based Perceptions of Long-Term Risk: Why Global Warming Does Not Scare Us (Yet)," *Climate Change* 70 (2006): 103–120.

19. As was the case in the 2002 survey, the dependent variable in the 2007, 2008, and 2011 surveys is measured using a survey item that was part of a framing experiment in which respondents received different information prior to expressing their preference about the future use of each energy sources. The analysis in this chapter includes only those respondents that were in the control groups in the experiments—that is, those individuals that did not receive any information. We discuss the experiments in detail in chapter 6.

20. Riley E. Dunlap and Aaron M. McCright, "A Widening Gap: Republican and Democratic Views on Climate Change," *Environment: Science and Policy for Sustainable Development* 50, no. 5 (2008): 26–35; and Aaron M. McCright and Riley E. Dunlap, "The Politicization of Climate Change and Polarization in the American Public's Views of Global Warming, 2001–2010," *The Sociological Quarterly* 52, no. 2 (2011): 155–194.

21. Nick F. Pidgeon, Irene Lorenzoni, and Wouter Poortinga, "Climate Change or Nuclear Power—No Thanks! A Quantitative Study of Public Perceptions and Risk Framing in Britain," *Global Environmental Change* 18, no. 1 (2008): 69–85.

22. Patrick Moore, "Going Nuclear," *Washington Post*, April 16, 2006, http://www.washingtonpost.com/wp-dyn/content/article/2006/04/14/AR2006041401209.html (accessed August 27, 2013).

23. Erika Lovely, "Environmentalists See Fission on Nuclear Power," January 31, 2008, http://www.politico.com/news/stories/0108/8241.html (accessed August 27, 2013).

24. Theda Skocpol, "Naming the Problem: What It Will Take to Counter Extremism and Engage Americans in the Fight Against Global Warming?," report prepared for the Symposium on the Politics of America's Fight Against Global Warming, January 2013.

25. See, for example, Robert Keohane and David Victor, "The Transnational Politics of Energy" *Daedalus* 142, no. 1 (2013): 97–109; and Krosnick and MacInnis, "Does the American Public Support Legislation to Reduce Greenhouse Gas Emissions?"

26. One way to calculate this is by taking the average absolute deviation from of column 2 from column 1 divided by the total deviation in the baseline (the sum of column 1). The reduction in total deviation is only 10 percent. That is, $[|1.87 - 1.79| + |1.58 - 1.43| + |1.24 - 1.03| + |0.91 - 0.75| + |0.67 - 0.62| + |0.04 - 0.02|)/(1.87 + 1.58 + 1.24 + 0.91 + 0.67 + 0.04) = 0.106$.

27. For a far-ranging discussion of these questions, see Smith, *Energy, the Environment, and Public Opinion.*

6 The Chicken and the Egg

1. For a thorough survey of such work, we refer the reader to Brian Gaines, Jim Kuklinski, and Paul Quirk, "The Logic of Survey Experimentation," *Political Analysis* 15, no. 1 (2007): 1–20. See also Jason Barabas and Jennifer Jerit, "Are Survey Experiments Externally Valid?," *American Political Science Review* 104, no. 2 (2010): 226–242.

2. The energy use preferences analyzed to this point in the book, for example in chapters 3 and 5, has focused on this group.

3. Because the surveys were done online, the random assignment was performed by the computer administering the survey and at the moment the respondent began the survey. The assignment of respondents to receive a particular sort of information is purely random: it is unrelated to any demographic facts about the respondents, the attitudes they may hold, or even their desires or interests to learn additional information about electricity generation. For this reason, we are able to infer the effect of additional information about electricity generation on a typical person or in the public as a whole.

4. MIT, "The Future of Nuclear Power."

5. It is important to note that the information included in the experiments was based on estimates of cost at the time. Over the decade of time covered in the experiments, the cost of generating electricity from different fuel sources obviously changed, in some cases (e.g., natural gas, wind) substantially. Given that the main purpose of the experiments is to examine the effects of changing the relative cost of different fuel types, using exactly the right price is not essential.

6. Gallup, "Environment," http://www.gallup.com/poll/1615/environment.aspx (accessed September 18, 2013).

7. MIT, "The Future of Coal: Options for a Carbon-Constrained World," Cambridge, MA, 2007.

8. People seem to overestimate the amount of oil used to generate electricity; currently, only about 1 percent of U.S. electricity generation comes from burning oil.

9. For this reason, we do not highlight the obvious and important hazardous waste generated from using nuclear power to generate electricity.

10. Michael Greenstone and Adam Looney, "Paying Too Much for Energy? The True Costs of Our Energy Choices," *Daedalus* 141, no. 2 (2012): 10–30. See also Greenstone and Looney, "A Strategy for America's Energy Future."

7 Two Minds about Climate Change

1. Daniel Yergin, *The Prize: The Epic Quest for Oil, Money, and Power* (New York: Simon and Shuster, 2008); Roy Licklider, "The Power of Oil: The Arab Oil Weapon and the Netherlands, the United Kingdom, Canada, Japan, and the United States," *International Studies Quarterly* 32 (1988): 205–226.

2. Alan S. Binder, "The Anatomy of Double-Digit Inflation in the 1970s," in *Inflation: Causes and Effects*, ed. Robert E. Hall, 261–282 (Chicago: University of Chicago Press, 1982); James D. Hamilton, "Oil and the Macroeconomy since World War II," *Journal of Political Economy* 91 (1983): 228, and Kevin L. Kliesen, "Rising Oil Prices and Economic Turmoil," *The Regional Economist*, Federal Reserve Bank of Saint Louis, January 2001. http://www.stlouisfed.org/publications/re/articles/?id=475 (accessed December 20, 2013). For a skeptical view of the effects of the oil shocks on the macroeconomy, see Robert B. Barsky and Killian Lutz, "Oil and the Macroeconomy since the 1970s," *The Journal of Economic Perspectives* 18 (2004): 115–134.

3. National Climatic Data Center, "Billion Dollar U.S. Weather/Climate Disasters," http://www.ncdc.noaa.gov/billions/ (accessed December 16, 2013).

4. James E. Hansen, "The Greenhouse Effect: Impacts on Current Global Temperature and Regional Heat Waves," testimony before U.S. Senate Committee on Energy and Natural Resources, June 23, 1988.

5. Eric Pooley, *The Climate War: True Believers, Power Brokers, and the Fight to Save the Earth* (New York: Hyperion, 2010).

6. Some research finds that whether the issue is framed as "global warming" or "climate change" can make a difference for some people in terms of their degree of concern. One recent experimental analysis found that, while Democrats were unaffected by question wording, Republicans were much more likely to endorse the issue as a real problem when framed as climate change rather than global warming. See Jonathon P. Schuldt, Sara H. Konrath, and Norbert Schwartz, "'Global Warming' or 'Climate Change'? Whether the Planet Is Warming Depends

on Question Wording," *Public Opinion Quarterly* 75, no. 1 (2011): 115–124. For a similar study, see Ann Villar and Jon A. Krosnick, "Global Warming vs. Climate Change, Taxes vs. Price: Does Word Choice," *Climatic Change* 105 (2011): 1–12.

7. Matthew E. Kahn and Matthew J. Kotchen, "Environmental Concern and the Business Cycle: The Chilling Effect of Recession," NBER Working Paper No. 16241, July 2010.

8. Lyle Scruggs and Salil Benegal, "Declining Public Concern about Climate Change: Can We Blame the Great Recession?" *Global Environmental Change* 22, no. 2 (2012): 505–515.

9. Leiserowitz et al., "Global Warming's Six Americas 2009."

10. Leiserowitz et al., "Global Warming's Six Americas, March 2012 & Nov. 2011."

11. See, for example, Christopher P. Borick and Barry G. Rabe, "A Reason to Believe: Examining the Factors that Determine Individual Views on Global Warming," *Social Science Quarterly* 91, no. 3 (2010): 777–800; Lawrence C. Hamilton, "Education, Politics and Opinions about Climate Change Evidence for Interaction Effects," *Climatic Change* 104, no. 2 (2011): 231–242; and Jon A. Krosnick, Allyson L. Holbrook, and Penny S. Visser, "The Impact of the Fall 2007 Debate about Global Warming on American Public Opinion," *Public Understanding of Science* 9 (2000): 239–260.

12. See, for example, Dunlap and McCright, "A Widening Gap"; and McCright and Dunlap, "The Politicization of Climate Change and the Polarization in the American Public's Views of Global Warming, 2001–2010."

13. A recent study of aggregate U.S. public concern for climate change found that it was best explained by elite cues and economic factors, and to a lesser extent media coverage. See Robert J. Brulle, Jason Carmichael, and J. Craig Jenkins, "Shifting Public Opinion on Climate Change: An Empirical Assessment of Factors Influencing Concern over Climate Change in the U.S., 2002–2010," *Climatic Change* 114, no. 2 (2012): 169–188. Other research similarly finds a large role for issue framing: Matthew C. Nisbet, "Communicating Climate Change: Why Frames Matter for Public Engagement," *Environment: Science and Policy for Sustainable Development* 51, no. 2 (2009): 12–23.

14. Anthony Leiserowitz, Edward W. Maibach, Connie Roser-Renouf, Nicholas Smith, and Erica Dawson, "Climategate, Public Opinion, and the Loss of Trust, July 2, 2010, http://papers.ssrn.com/sol3/papers.cfm?abstract_id=1633932 (accessed December 16, 2013); and Chris Mooney, "The Science of Why We Don't Believe Science," *Mother Jones*, April 18, 2011.

15. The OLS regression model treats the ordinal dependent variable as continuous. We also estimated ordinal logistic regression models with identical substantive results, but opted to present and discuss the OLS estimates for clarity.

16. Specifically, Gallup grouped a number of economy-related responses, ranging from the economy in general and unemployment to income inequality and wage issues.

17. Gallup, "U.S. Satisfaction Up Slightly at Start of 2012, to 18%," January 11, 2012, http://www.gallup.com/poll/1675/Most-Important-Problem.aspx (accessed August 27, 2013).

18. This analysis is based on data compiled as part of the Policy Agendas project. For more information, see the "Trends Analysis" at http://www.policyagendas.org (accessed August 27, 2013).

19. This question design differs from that used by Gallup, which is an open-ended item in which respondents volunteer a problem.

20. For more detailed analysis of these data, see Konisky, Milyo, and Richardson, "Environmental Policy Attitudes"; and David M. Konisky "Public Preferences for Environmental Policy Responsibility," *Publius: The Journal of Federalism* 41, no. 1 (2011): 76–100.

21. To develop these rankings, we coded a response of "A lot more" as a 2, "A little bit more" as 1, "About the same" as 0, "A little bit less" as -1, and "A lot less" as -2. Positive values therefore indicate a preference for more government effort to address the issue, while negative values reflect a preference for less government intervention.

22. For a discussion of the willingness-to-pay concept, see Nathaniel O. Keohane and Sheila Olmstead, *Markets and the Environment* (Washington, DC: Island Press, 2007).

23. For a good debate about the use of contingent valuation, refer to a symposium in the 1994 issue of *The Journal of Economic Perspectives*.

24. The 2013 survey was designed slightly differently than the previous surveys in that the starting value was set to $15, rather than $5. The data suggest a tendency of some segment of the population to say yes to the first value, whatever it is, which is reflected in the 30 percent of the respondents that indicated that they would be willing pay $15. As a consequence, mean and median values for the 2013 may be overinflated.

25. Evan Johnson and Gregory F. Nemet, "Willingness to Pay for Climate Policy: A Review of Estimates," La Follette School Working Paper No. 2010–011, Madison, WI, June 2010.

26. Congressional Budget Office, "The Estimated Costs to Households From the Cap-and-Trade Provisions to H.R. 2454.

27. Beach et al., "Son of Waxman-Markey."

8 What to Do?

1. Interestingly, although California is the second largest electricity consumer, it has the lowest per capita electricity consumption in the United States. That fact was both a point of pride and a cause for concern in the discussion of Proposition 23, as the further efficiency gains would likely be expensive. Energy Information Administration, "State Profiles and Energy Estimates," January 16, 2014, http://www.eia.gov/state (accessed January 21, 2014).

2. Proposition 7 required a Renewable Portfolio Standard of 40 percent alternative fuels by 2020 and 50 percent by 2025. The law would have been so stringent that even the Sierra Club opposed it. Proposition 10 proposed bonds to fund construction of wind and other energy infrastructure; it was strongly supported and financed by T. Boone Pickens.

3. Steven Mufson, "Calif.'s Prop 23 Battle Pits Big Oil against Environmental Concerns," *Washington Post*, October 21, 2010, http://www.washingtonpost.com/wp-dyn/content/article/2010/10/21/AR2010102102967.html (accessed December 20, 2013).

4. "Proposition 23 Poll Shows a Dead Heat among California Voters," *Los Angeles Times*, September 24, 2010, http://latimesblogs.latimes.com/greenspace/2010/09/proposition-23-poll-global-warming-california.html (accessed December 20, 2013).

5. For a detailed discussion, see Joseph E. Aldy and Robert N. Stavins, "Using the Market to Address Climate Change: Insights from Theory and Experience," *Daedalus* 141, no. 2 (2012): 45–60.

6. The EPA's effort to regulate mercury and other air toxics actually began in the mid-1990s, when the agency initiated a study of these pollutants. In December, 2000, the EPA announced that it was "appropriate and necessary" to regulate coal and oil power plants to reduce emissions of such toxic substances, and that decision initiated the rule making process. See "Regulatory Actions | Mercury and Air Toxics Standards (MATS) for Power Plants," June 25, 2013, http://www.epa.gov/mats/actions.html (accessed December 14, 2013).

7. Felicia Sonmez, "REINS Bill to Expand Congressional Power over Executive Regulations Passed by House," *Washington Post*, December 7, 2011, http://www.washingtonpost.com/blogs/2chambers/post/reins-bill-to-expand-congressional-power-over-executive-regulations-passed-by-house/2011/12/07/gIQAs6VMdO_blog.html (accessed December 14, 2013).

8. In June 2012, the United States Court of Appeals for the District of Columbia held in the case *Coalition for Responsible Regulation, Inc. et al. v. Environmental Protection Agency* that EPA's greenhouse gas regulations were justified under the CAA.

9. For an excellent discussion of the effectiveness of the SO_2 system, see Richard Schmalensee and Robert Stavins, "The SO_2 Allowance Trading System: The Ironic History of a Grand Policy Experiment," *Journal of Economic Perspectives* 27, no. 1 (2013): 103–122. It also should be noted that the SO_2 market has recently collapsed as a result of a series of court and regulatory decisions.

10. European Union, "The EU Emissions Trading System (EU ETS), January 8, 2014, http://ec.europa.eu/clima/policies/ets/index_en.htm (accessed January 21, 2014).

11. State of California, "California Climate Change Portal," http://www.climatechange.ca.gov/ (accessed January 21, 2014).

12. Ministry of Finance, "British Columbia, Budget and Fiscal Plan 2008/09–2010/11," February 19, 2008.

13. Andrea Campbell, "What Americans Think of Taxes," in *The New Fiscal Sociology*, ed. Monica Prasad, Isaac Martin, and Ajay Mehrotra, 48–67 (New York: Cambridge University Press, 2009).

14. The surveys analyzed here include March 2006 and March 2012 surveys by Gallup; January 2007, June 2008, April 2009, October 2009, and April 2001 surveys by CNN/ORC; April 2006 and April 2007 surveys by *ABC News/Washington Post*/Stanford; April 2009, June 2009, December 2009, and June 2010 surveys by *ABC News/Washington Post*; July 2008, June 2009, August 2009, and November 2009 surveys by *ABC News/Washington Post*; April 2009 and October 2009 surveys by *NBC News/Wall Street Journal*; a December 2009 survey from Ipsos/McClatchy; a June 2010 survey by *USA Today*/Gallup; and an August 2012 survey by the *Washington Post*/Kaiser.

15. In the 2010 survey, support for cap and trade increased to about 40 percent when paired with a tax rebate to consumers, but support for a carbon tax actually went down a couple of points when paired with a similar tax rebate.

16. Jeffrey L. Jordan and Abdelmoneim H. Elnagheeb, "Willingness to Pay for Improvements in Drinking Water Quality," *Water Resources Research* 29, no. 2 (1993): 237–245.

17. Diane Hite, Darren Hudson, and Walaiporn Intarapapong, "Willingness to Pay for Water Quality Improvements," *Journal of Agricultural and Resource Economics* 27, no. 2 (2002): 433–449.

18. F. W. Gramlich, "The Demand for Clean Water: The case of the Charles River," *National Tax Journal* 30, no. 2 (1977): 183–194.

19. Jason F. Shogren, *The Benefits and Costs of the Kyoto Protocol* (Washington, DC: American Enterprise Institute, 1999).

20. These results have appeared in various issues of the *Boston Review* over the past few years.

21. See, for example, Rob Stavins's blog: http://www.robertstavinsblog.org/. This is a great resource for anyone who follows the ongoing political developments in climate policy and current economic thinking on the subject.

22. In the case of the straight carbon tax, the responses were on a three-point scale, which we coded as 1 = Support, 0.5 = Neither support nor oppose and 0 = Oppose).

23. The variables are measured using the rotated factor scores from a principal component factor analysis. Factor analysis attempts to measure the dimensions underlying a set of variables. That is, suppose a set of variables—in our case, say, opinions about costs of six different energy sources—are really a function of a couple of dimensions. The correlations among people's answers will reflect those two dimensions, with, say, understanding of the costs of traditional fuels, with oil being regarded as expensive and coal as cheap among the fossil fuels, and also attitudes about costs of fossil fuels versus the costs of alternative energy, with some people believing that alternative energy is expensive and traditional fuels cheap and others thinking that alternative energy is cheap and traditional fuels relatively expensive. Factor analysis reveals the dimensions of the data from the

structure of the correlations. It also indicates how each variable translates into each dimension, using a set of weights called factor loadings. The factor analysis of the harm and of the cost variables for nuclear, coal, oil, natural gas, hydroelectricity, wind, and solar power revealed two dimensions for both harms and costs. Opinions about the costs or harms of coal, natural gas, nuclear, oil, and hydro load onto one dimension, and opinions about harms and costs of solar and wind power onto a second dimension.

24. We use OLS regression for simplicity of presentation, so these can be thought of as linear probability models.

25. It could also reflect measurement error, such as that which arises from vague or confusing questions.

26. Barry G. Rabe,"States on Steroids: The Intergovernmental Odyssey of American Climate Policy," *Review of Policy Research* 25, no. 2 (2008): 105–128.

9 A Way Forward

1. Carmines and Stimson, *Issue Evolution*; Page and Shapiro, *The Rational Public*; and Robert S. Erickson, Michael B. MacKuen, and James A. Stimson, *The Macro Polity* (New York: Cambridge University Press, 2002).

2. Graetz, *The End of Energy*.

3. Greenstone and Looney, "A Strategy for America's Energy Future."

4. Ibid., 15.

5. Greenstone and Looney, "Paying Too Much for Energy?"

6. MIT, "The Future of Coal."

Bibliography

Ackerman, Frank, and Lisa Heinzerling. *Priceless: On Knowing the Price of Everything and the Value of Nothing*. New York: The New Press, 2004.

Aldy, Joseph E. "Promoting Clean Energy in the American Power Sector." The Hamilton Project, Washington, DC, May 2011.

Aldy, Joseph E., and Robert N. Stavins. "Using the Market to Address Climate Change: Insights from Theory and Experience." *Daedalus* 141, no. 2 (2012): 45–60.

Ansolabehere, Stephen, and David M. Konisky. "Public Attitudes toward Construction of New Power Plants." *Public Opinion Quarterly* 73, no. 3 (2009): 566–577.

Ansolabehere, Stephen, and Brian Schaffner. "Does Survey Mode Still Matter? Findings from a 1010 Multi-Mode Comparison." Forthcoming in *Political Analysis*.

Ansolabehere, Stephen, Marc Meredith, and Erik Snowberg. "Asking about Numbers: Why and How." *Political Analysis* 21, no. 1 (2013): 48–69.

Barabas, Jason, and Jennifer Jerit. "Are Survey Experiments Externally Valid?" *American Political Science Review* 104, no. 2 (2010): 226–242.

Barsky, Robert B., and Killian Lutz. "Oil and the Macroeconomy since the 1970s." *Journal of Economic Perspectives* 18 (2004): 115–134.

Baumgartner, Frank R., and Bryan D. Jones. *Agendas and Instability in American Politics*. Chicago: University of Chicago Press, 1993.

Beach, William W., David Kreutzer, Karen Campbell, and Ben Lieberman. "Son of Waxman-Markey: More Politics Makes for a More Costly Bill." Heritage Foundation WebMemo No. 2450, May 18, 2009.

Bidwell, David. "The Role of Values in Public Beliefs and Attitudes towards Commercial Wind Energy." *Energy Policy* 58 (2013): 189–199.

Binder, Alan S. "The Anatomy of Double-Digit Inflation in the 1970s." In *Inflation: Causes and Effects*, ed. Robert E. Hall, 261–282. Chicago: University of Chicago Press, 1982.

Bolsen, Toby, and Fay Lomax Cook. "The Polls—Trends: Public Opinion on Energy Policy: 1974–2006." *Public Opinion Quarterly* 72 (2008): 364–388.

Borick, Christopher P., and Barry G. Rabe. "A Reason to Believe: Examining the Factors that Determine Individual Views on Global Warming." *Social Science Quarterly* 91, no. 3 (2010): 777–800.

Broder, John. "Nebraska Governor Approves Keystone XL Route." *New York Times*, January 22, 2013.

Brulle, Robert J., Jason Carmichael, and J. Craig Jenkins. "Shifting Public Opinion on Climate Change: An Empirical Assessment of Factors Influencing Concern over Climate Change in the U.S., 2002–2010." *Climatic Change* 114, no. 2 (2012): 169–188.

Campbell, Andrea. "What Americans Think about Taxes." In *The New Fiscal Sociology*, ed. Monica Prasad, Isaac Martin, and Ajay Mehrotra, 48–67. New York: Cambridge University Press, 2009.

Carman, Christopher Jan. 1998. "Dimensions of Environmental Policy Support in the United States." *Social Science Quarterly* 79, no. 4: 717–733.

Carmines, Ted, and James Stimson. *Issue Evolution*. Princeton: Princeton University Press, 1989.

Congressional Budget Office. "The Estimated Costs to Households from the Cap-and-Trade Provisions of H.R. 2454." June 19, 2009. http://www.cbo.gov/sites/default/files/cbofiles/ftpdocs/103xx/doc10327/06-19-capandtradecosts.pdf (accessed August 17, 2013).

Davis, Lucas. "The Effect of Power Plants on Local Housing Prices and Rents: Evidence from Restricted Census Microdata." *Review of Economics and Statistics* 93, no. 4 (2011): 1391–1402.

Deutch, John M. *The Crisis in Energy Policy*. Cambridge, MA: Harvard University Press, 2001.

Dietz, Thomas, Amy Dan, and Rachael Shwom. "Support for Climate Change Policy: Social Psychological and Social Structural Influences." *Rural Sociology* 72, no. 2 (2007): 185–214.

Dietz, Thomas, Paul C. Stern, and Elke U. Weber. "Reducing Carbon-Based Energy Consumption through Changes in Household Behavior." *Daedalus* 142 (Winter 2013): 78–89.

Douglas, Mary, and Aaron Wildavsky. *Risk and Culture*. Berkeley: University of California Press, 1992.

Du, Yangbo, and John E. Parsons. "Update on the Costs of Nuclear Power." Working Paper 09-994, MIT Center for Energy and Environmental Policy Research, May 2009.

Duke Energy. "How Do Coal-Fired Plants Work?" http://www.duke-energy.com/about-energy/generating-electricity/coal-fired-how.asp (accessed August 26, 2013).

Dunlap, Riley E., and Aaron M. McCright.. "A Widening Gap: Republican and Democratic Views on Climate Change." *Environment: Science and Policy for Sustainable Development* 50, no. 5 (2008): 26–35.

Dunlap, Riley E., Chenyang Xiao, and Aaron M. McCright. "Politics and Environment in America: Partisan and Ideological Cleavages in Public Support for Environmentalism." *Environmental Politics* 10, no. 4 (2001): 23–48.

Erikson, Robert, Michael B. MacKuen, and James A. Stimson. *The Macro Policy*. New York: Cambridge University Press, 2002.

European Union. "The EU Emissions Trading System (EU ETS)." January 8, 2014. http://ec.europa.eu/clima/policies/ets/index_en.htm (accessed January 21, 2014).

Farhar, Barbara C. "Trends in U.S. Public Perceptions and Preferences on Energy and Environmental Policy." *Annual Review of Energy and the Environment* 19 (1994): 211–239.

Farhar, Barbara C., Charles T. Unseld, Rebecca Vories, and Robin Crews. "Public Opinion about Energy." *Annual Review of Energy* 5 (1980): 141–172.

Firestone, Jeremy, and Willett Kempton. "Public Opinion about Large Offshore Wind Power: Underlying Factors." *Energy Policy* 35, no. 3 (2007): 1584–1598.

Fishkin, J. "Deliberative Polling: Toward a Better-Informed Democracy." Center for Deliberative Democracy. 2010. cdd.stanford.edu/polls/docs/summary (accessed December 20, 2013).

Freudenburg, William A., and Eugene A. Rosa. *Public Reaction to Nuclear Power: Are There Critical Masses?* Boulder, CO: Westview Press, 1984.

"The Frontier Spirit." *Economist*, May 14, 1998. http://www.economist.com/node/128792.

Fullerton, Don, and Catherine Wolfram, eds. *The Design and Implementation of U.S. Climate Policy*. Chicago: University of Chicago Press, 2012.

Gaines, Brian, Jim Kuklinski, and Paul Quirk. "The Logic of Survey Experimentation." *Political Analysis* 15, no. 1 (2007): 1–20.

Gallagher, Kelly Sims. *Acting in Time on Energy Policy*. Washington, DC: Brookings Institution Press, 2009.

Gallup. "Americans Still Divided on Energy-Environment Trade-Off." April 10, 2013. http://www.gallup.com/poll/161729/americans-divided-energy-environment-trade-off.aspx (accessed September 18, 2013).

Gallup. "Environment." http://www.gallup.com/poll/1615/environment.aspx (accessed September 18, 2013).

Gallup. "More Americans Still Prioritize Economy over Environment." April 3, 2013. http://www.gallup.com/poll/161594/americans-prioritize-economy-environment.aspx (accessed August 27, 2013).

Gallup. "U.S. Satisfaction Up Slightly at Start of 2012, to 18%." January 11, 2012. http://www.gallup.com/poll/1675/Most-Important-Problem.aspx (accessed August 27, 2013).

Graetz, Michael J. *The End of Energy: The Unmaking of America's Environment, Security, and Independence*. Cambridge, MA: MIT Press, 2011.

Gramlich, F. W. "The Demand for Clean Water: The Case of the Charles River." *National Tax Journal* 30, no. 2 (1977): 183–194.

Greenstone, Michael, and Adam Looney. "A Strategy for America's Energy Future: Illuminating Energy's Full Costs." The Hamilton Project, The Brookings Institution, Washington, DC, 2011.

Greenstone, Michael, and Adam Looney. "Paying Too Much for Energy? The True Costs of Our Energy Choices." *Daedalus* 141, no. 2 (2012): 10–30.

Guber, Deborah Lynn. *The Grassroots of a Green Revolution: Polling America on the Environment*. Cambridge, MA: MIT Press, 2003.

Haber, Deborah. "Seventies Oil Crisis Was a 'Perfect Storm' for U.S." MIT News, March 23, 2007. http://newsoffice.mit.edu/2007/seventies-oil-crisis-was-perfect-storm-us (accessed April 30, 2014).

Hamilton, James D. "Oil and the Macroeconomy since World War II." *Journal of Political Economy* 91 (1983): 228.

Hamilton, Lawrence C. "Education, Politics and Opinions about Climate Change Evidence for Interaction Effects." *Climatic Change* 104, no. 2 (2011): 231–242.

Hansen, James E. "The Greenhouse Effect: Impacts on Current Global Temperature and Regional Heat Waves." Testimony before U.S. Senate Committee on Energy and Natural Resources, June 23, 1988.

Hite, Diane, Darren Hudson, and Walaiporn Intarapapong. "Willingness to Pay for Water Quality Improvements." *Journal of Agricultural and Resource Economics* 27, no. 2 (2002): 433–449.

Hunter, Susan, and Kevin M. Leyden. "Beyond NIMBY: Explaining Opposition to Hazardous Waste Facilities." *Policy Studies Journal: The Journal of the Policy Studies Organization* 23 (1995): 601–619.

International Monetary Fund. World Economic Outlook Database. October 2010. http://www.imf.org/external/pubs/ft/weo/2010/02/weodata/index.aspx (accessed December 20, 2013).

Iyengar, Shanto, and Donald R. Kinder. *News that Matters*. Chicago: University of Chicago Press, 1989.

Jenkins-Smith, Hank, and Howard Kunreuther. "Mitigation and Benefits Measures as Policy Tools for Siting Potentially Hazardous Facilities: Determinants of Effectiveness and Appropriateness." *Risk Analysis* 21 (2001): 371–382.

Johnson, Evan, and Gregory F. Nemet. "Willingness to Pay for Climate Policy: A Review of Estimates." La Follette School Working Paper No. 2010-011, Madison, WI, June 2010.

Joppke, Christian. *Mobilizing against Nuclear Energy: A Comparison of Germany and the United States*. Berkeley: University of California Press, 1993.

Jordan, Jeffrey L., and Abdelmoneim H. Elnagheeb. "Willingness to Pay for Improvements in Drinking Water Quality." *Water Resources Research* 29, no. 2 (1993): 237–245.

Joskow, Paul. "Energy Policies and Their Consequences 25 Years Later." *Energy Journal* 24, no. 4 (2003): 17–49.

Joskow, Paul L., and Richard Schmalensee. *Markets for Power*. Cambridge, MA: MIT Press, 1988.

Kahn, Matthew E., and Matthew J. Kotchen. "Environmental Concern and the Business Cycle: The Chilling Effect of Recession." NBER Working Paper No. 16241, July 2010.

Kamieniecki, Sheldon. "Political Parties and Environmental Policy." In *Environmental Politics and Policy*, 2nd ed., ed. James P. Lester, 146–167. Durham, NC: Duke University Press, 1995.

Keohane, Nathaniel O., and Sheila Olmstead. *Markets and the Environment*. Washington, DC: Island Press, 2007.

Keohane, Robert, and David Victor. "The Transnational Politics of Energy." *Daedalus* 142, no. 1 (2013): 97–109.

Kliesen, Kevin L. "Rising Oil Prices and Economic Turmoil." *The Regional Economist*, Federal Reserve Bank of St. Louis, January 2001. http://www.stlouisfed.org/publications/re/articles/?id=475 (accessed December 20, 2013).

Klineberg, Stephen L., Matthew McKeever, and Bert Rothenbach. "Demographic Predictors of Environmental Concern: It Does Make a Difference How It's Measured." *Social Science Quarterly* 79, no. 4 (1998): 734–753.

Konisky, David M. "Public Preferences for Environmental Policy Responsibility." *Publius: The Journal of Federalism* 41, no. 1 (2011): 76–100.

Konisky, David M., Jeffrey Milyo, and Lilliard E. Richardson Jr. "Environmental Policy Attitudes: Issues, Geographical Scale, and Political Trust." *Social Science Quarterly* 89, no. 5 (2008): 1066–1085.

Krauss, Clifford. "Move Over, Oil, There's Money in Texas Wind." *New York Times*, February 23, 2008. http://www.nytimes.com/2008/02/23/business/23wind.html?_r=2&hp&oref=slogin&

Krohn, Søren, and Steffen Damborg. "On Public Attitudes towards Wind Power." *Renewable Energy* 16, no. 1 (1999): 954–960.

Krosnick, Jon A. "The Climate Majority." *New York Times*, June 8, 2010.

Krosnick, Jon A., and Bo MacInnis. "Does The American Public Support Legislation to Reduce Greenhouse Gas Emissions?" *Daedalus* 142, no. 1 (2013): 26–39.

Krosnick, Jon A., Allyson L. Holbrook, and Penny S. Visser. "The Impact of the Fall 2007 Debate about Global Warming on American Public Opinion." *Public Understanding of Science* 9 (2000): 239–260.

Layzer, Judy. *The Environmental Case*. 3rd ed. Washington, DC: CQ Press, 2011.

Lehr, R. L., W. Guild, D. L. Thomas, and B. G. Swezey. "Listening to Customers: How Deliberative Polling Helped Build 1,000 MB of New Renewable Energy Projects in Texas." Technical Report, NREL/TP-620-33177. National Renewable Electricity Laboratory, Golden, CO, June 2003.

Leiby, Paul N. "Estimating the Energy Security Benefits of Reduced U.S. Oil Imports." Final Report, ORNL/TM-2007/028, Oak Ridge National Laboratory, revised March 14, 2008.

Leiserowitz, Anthony. "Climate Change Risk Perception and Policy Preferences: The Role of Affect, Imagery, and Values." *Climatic Change* 77, nos. 1–2 (2006): 45–72.

Leiserowitz, A., E. Maibach, C. Roser-Renouf, and J. Hmielowski. "Global Warming's Six Americas 2009: An Audience Segmentation Analysis, May 2009." Yale University and George Mason University. Yale Project on Climate Change Communication, New Haven, CT, 2012.

Leiserowitz, A., E. Maibach, C. Roser-Renouf, and J. Hmielowski. "Global Warming's Six Americas, March 2012 & Nov. 2011." Yale University and George Mason University, Yale Project on Climate Change Communication, New Haven, CT, 2012.

Leiserowitz, Anthony, Edward W. Maibach, Connie Roser-Renouf, Nicholas Smith, and Erica Dawson. "Climategate, Public Opinion, and the Loss of Trust." *American Behavioral Scientist* 57 (2012): 818–837.

Lester, Richard K., and David M. Hart. *Unlocking Energy Innovation: How America Can Build a Low-Cost, Low-Carbon Energy System*. Cambridge, MA: MIT Press, 2012.

Licklider, Roy. "The Power of Oil: The Arab Oil Weapon and the Netherlands, the United Kingdom, Canada, Japan, and the United States." *International Studies Quarterly* 32 (1988): 205–226.

Lovely, Erika. "Environmentalists See Fission on Nuclear Power." January 31, 2008. http://www.politico.com/news/stories/0108/8241.html (accessed August 27, 2013).

McCright, Aaron, and Riley E. Dunlap. "The Politicization of Climate Change and Polarization in the American Public's Views of Global Warming, 2001–2010." *Sociological Quarterly* 52, no. 2 (2011): 155–194.

Ministry of Finance. "British Columbia, Budget and Fiscal Plan 2008/09–2010/11." February 19, 2008.

MIT. "The Future of Nuclear Power." Cambridge, MA, 2003.

MIT. "The Future of Coal: Options for a Carbon-Constrained World." Cambridge, MA, 2007.

Moniz, Ernest. "Why We Still Need Nuclear Power: Making Clean Energy Safe and Affordable." *Foreign Affairs*, November/December 2011.

Mooney, Chris. "The Science of Why We Don't Believe Science." *Mother Jones*, April 18, 2011.

Moore, Patrick. "Going Nuclear." *Washington Post*, April 16, 2006. http://www.washingtonpost.com/wp-dyn/content/article/2006/04/14/AR2006041401209.html (accessed August 27, 2013).

Mufson, Steven. "Calif.'s Prop 23 Battle Pits Big Oil against Environmental Concerns." *Washington Post*, October 21, 2010. http://www.washingtonpost

.com/wp-dyn/content/article/2010/10/21/AR2010102102967.html (accessed December 20, 2013).

National Climatic Data Center. "Billion Dollar U.S. Weather/Climate Disasters." http://www.ncdc.noaa.gov/billions/ (accessed December 16, 2013).

National Research Council. 2010. *Hidden Costs of Energy*. Washington, DC: National Academies Press.

Nilsson, Andreas, Chris von Borgstede, and Anders Biel. "Willingness to Accept Climate Change Strategies: The Effect of Values and Norms." *Journal of Environmental Psychology* 24, no. 3 (2004): 267–277.

Nisbet, Matthew C. "Communicating Climate Change: Why Frames Matter for Public Engagement." *Environment: Science and Policy for Sustainable Development* 51, no. 2 (2009): 12–23.

Nisbet, Matthew C., and Teresa Myers. "The Polls—Trends Twenty Years of Public Opinion about Global Warming." *Public Opinion Quarterly* 71, no. 3 (2007): 444–470.

Nuclear Energy Institute. "Nuclear Waste: Amounts & On-Site Storage." http://www.nei.org/Knowledge-Center/Nuclear-Statistics/On-Site-Storage-of-Nuclear-Waste (accessed December 16, 2013).

O'Connor, Robert E., Richard J. Bord, and Ann Fisher. "Risk Perceptions, General Environmental Beliefs, and Willingness to Address Climate Change." *Risk Analysis* 19, no. 3 (1999): 461–471.

Page, Benjamin, and Robert Shapiro. *The Rational Public: Fifty Years of Trends in Americans' Policy Preferences*. Chicago: University of Chicago Press, 1992.

Pearce, David. "The Social Cost of Carbon and its Policy Implications." *Oxford Review of Economic Policy* 19, no. 3 (2003): 362–384.

Pidgeon, Nick F., Irene Lorenzoni, and Wouter Poortinga. "Climate Change or Nuclear Power—No Thanks! A Quantitative Study of Public Perceptions and Risk Framing in Britain." *Global Environmental Change* 18, no. 1 (2008): 69–85.

Pollock, Philip H., III, M. Elliot Vittes, and Stuart A. Lilie. "Who Says It's Risky Business? Public Attitudes toward Hazardous Waste Facility Siting." *Polity* 24 (1992): 499–513.

Pooley, Eric. *The Climate War: True Believers, Power Brokers, and the Fight to Save the Earth*. New York: Hyperion, 2010.

Portney, Kent E. *Siting Hazardous Waste Treatment Facilities: The NIMBY Syndrome*. New York: Auburn House, 1991.

"Proposition 23 Poll Shows a Dead Heat among California Voters." *Los Angeles Times*, September 24, 2010. http://latimesblogs.latimes.com/greenspace/2010/09/proposition-23-poll-global-warming-california.html (accessed December 20, 2013).

Rabe, Barry G. "States on Steroids: The Intergovernmental Odyssey of American Climate Policy." *Review of Policy Research* 25, no. 2 (2008): 105–128.

"Regulatory Actions | Mercury and Air Toxics Standards (MATS) for Power Plants." June 25, 2013. http://www.epa.gov/mats/actions.html (accessed December 14, 2013).

Rivers, Douglas. "Sampling for Web Surveys." Paper presented at the 2007 Joint Statistical Meetings, Salt Lake City, UT, 2007.

Rosa, Eugene, and Riley Dunlap. "Nuclear Power: Three Decades of Public Opinion." *Public Opinion Quarterly* 58 (1994): 295–324.

Schmalensee, Richard, and Robert Stavins. "The SO_2 Allowance Trading System: The Ironic History of a Grand Policy Experiment." *Journal of Economic Perspectives* 27, no. 1 (2013): 103–122.

Schuldt, Jonathon P., Sara H. Konrath, and Norbert Schwartz. "'Global Warming' or 'Climate Change'? Whether the Planet Is Warming Depends on Question Wording." *Public Opinion Quarterly* 75, no. 1 (2011): 115–124.

Schuman, Howard, and Stanley Presser. "The Open and Closed Question." *American Sociological Review* 44 (October 1979): 692–712.

Scruggs, Lyle, and Salil Benegal. "Declining Public Concern about Climate Change: Can We Blame the Great Recession?" *Global Environmental Change* 22, no. 2 (2012): 505–515.

Shogren, Jason F. *The Benefits and Costs of the Kyoto Protocol.* Washington, DC: American Enterprise Institute, 1999.

Shwom, Rachael, David Bidwell, Amy Dan, and Thomas Dietz. "Understanding U.S. Public Support for Domestic Climate Change Policies." *Global Environmental Change* 20, no. 3 (2010): 472–482.

Skocpol, Theda. "Naming the Problem: What It Will Take to Counter Extremism and Engage Americans in the Fight against Global Warming?" Report prepared for the Symposium on the Politics of America's Fight against Global Warming, January 2013.

Slovic, Paul. "Perception of Risk." *Science* 236 (1987): 280–285.

Slovic, Paul. "Perception of Risk and the Future of Nuclear Power." *Arizona Journal of International and Comparative Law* 9 (1992): 191–198.

Slovic, Paul. "Perception of Risk and the Future of Nuclear Power." *Physics and Society* 23, no. 1 (1994): 3–5.

Slovic, Paul, James H. Flynn, and Mark Layman. "Perceived Risk, Trust, and the Politics of Nuclear Waste." *Science* 254, no. 5038 (1991): 1603–1607.

Smith, Eric R. A. N. *Energy, the Environment, and Public Opinion.* Lanham, MD: Rowman and Littlefield Publishers, 2003.

Sonmez, Felicia. "REINS Bill to Expand Congressional Power over Executive Regulations Passed by House." *Washington Post*, December 7, 2011. http://www.washingtonpost.com/blogs/2chambers/post/reins-bill-to-expand-congressional-power-over-executive-regulations-passed-by-house/2011/12/07/gIQAs6VMdO_blog.html (accessed December 14, 2013).

State of California. "California Climate Change Portal." http://www.climatechange.ca.gov/ (accessed January 21, 2014).

Stern, Nicholas. *The Economics of Climate Change: The Stern Review*. Cambridge: Cambridge University Press, 2006.

Tanaka, Yutaka. "Major Psychological Factors Determining Public Acceptance of the Siting of Nuclear Facilities." *Journal of Applied Social Psychology* 34 (2004): 1147–1165.

Tol, Richard S. J. "The Marginal Damage Costs of Carbon Dioxide Emissions: An Assessment of the Uncertainties." *Energy Policy* 33, no. 16 (2005): 2064–2074.

Tuttle, Robert, and Ola Galal. "Oil Ministers See Demand, Prices Rising, Undeterred by Greek Debt Crisis." *Bloomberg News*, May 10, 2010. http://www.bloomberg.com/news/2010-05-09/saudi-arabian-algerian-oil-ministers-see-consumption-increasing-this-year.html (accessed December 20, 2013).

U.S. Energy Information Administration. *Annual Energy Outlook 2012*. DOE/EIA-0383(2012). U.S. Department of Energy, Washington, DC, July 2012.

U.S Energy Information Administration. *Annual Energy Review 2009*. DOE/EIA-0384(2009). U.S. Department of Energy, Washington, DC, August 2010.

U.S. Energy Information Administration. *Electric Power Annual 2009*. DOE/EIA-0384(2009). U.S. Department of Energy, Washington, DC, January 2011.

U.S. Energy Information Administration. *Electric Power Annual 2011*. U.S. Department of Energy, Washington, DC, January 2013.

U.S. Energy Information Administration. "Projected Retirements of Coal-Fired Power Plants." July 31, 2012. http://www.eia.gov/todayinenergy/detail.cfm?id=7330 (accessed August 26, 2013).

U.S. Environmental Protection Agency. "Mercury and Air Toxics Standards." http://www.epa.gov/airquality/powerplanttoxics/ (accessed April 1, 2013).

U.S. Environmental Protection Agency. "EPA Analysis of the American Power Act in the 111th Congress." June 2010. http://www.epa.gov/climatechange/Downloads/EPAactivities/EPA_APA_Analysis_6-14-10.pdf (accessed August 17, 2013).

U.S. Environmental Protection Agency. "National Greenhouse Gas Emissions Data." 2011. http://www.epa.gov/climatechange/emissions/usinventoryreport.html (accessed December 20, 2013).

U.S. Nuclear Regulatory Commission. "Three Mile Island Accident." 2013. http://www.nrc.gov/reading-rm/doc-collections/fact-sheets/3mile-isle.pdf (accessed December 16, 2013).

Vavreck, Lynn, and Douglas Rivers. "The 2006 Cooperative Congressional Election Study." *Journal of Elections, Public Opinion and Parties* 18 (2008): 355–366.

Villar, Ann, and Jon A. Krosnick. "Global Warming vs. Climate Change, Taxes vs. Price: Does Word Choice." *Climatic Change* 105 (2011): 1–12.

Warren, Charles R., Carolyn Lumsden, Simone O'Dowd, and Richard V. Birnie. "'Green on Green': Public Perceptions of Wind Power in Scotland and Ireland." *Journal of Environmental Planning and Management* 48 (2005): 853–875.

Weber, Elke U. "Experience-Based and Description-Based Perceptions of Long-term Risk: Why Global Warming Does Not Scare Us (Yet)." *Climatic Change* 70 (2006): 103–120.

Weiss, Charles, and William B. Bonvillian. *Structuring and Energy Revolution.* Cambridge, MA: MIT Press, 2009.

Wolsink, Maarten. "Wind Power and the NIMBY-Myth: Institutional Capacity and the Limited Significance of Public Support." *Renewable Energy* 21, no. 1 (2000): 49–64.

Wolsink, Maarten. "Wind Power Implementation: The Nature of Public Attitudes: Equity and Fairness Instead of 'Backyard Motives.'" *Renewable & Sustainable Energy Reviews* 11 (2007): 1118–1207.

Worcester, Robert. "Public Opinion and the Environment." *The Political Quarterly* 68 (1997): 160–173.

Yergin, Daniel. *The Prize: The Epic Quest for Oil, Money, and Power.* New York: Simon and Shuster, 2008.

Zaller, John. *The Nature and Origin of Public Opinion.* Cambridge: Cambridge University Press, 1991.

Index